Automotive Audits

Practical Quality of the Future: What it Takes to be Best in Class (BIC)

Series Editor:

D. H. Stamatis, President of Contemporary Consultants, MI, USA

Quality Assurance
Applying Methodologies for Launching New Products, Services, and
Customer Satisfaction
D. H. Stamatis

Advanced Product Quality Planning
The Road to Success
D. H. Stamatis

Automotive Audits
Principles and Practices
D. H. Stamatis

Automotive Process Audits
Preparations and Tools
D. H. Stamatis

For more information about this series, please visit: https://www.routledge.com/
Practical-Quality-of-the-Future/book-series/PRAQUALFUT

Automotive Audits
Principles and Practices

D. H. Stamatis

CRC Press
Taylor & Francis Group
Boca Raton London New York

CRC Press is an imprint of the
Taylor & Francis Group, an **informa** business

First edition published 2021
by CRC Press
6000 Broken Sound Parkway NW, Suite 300, Boca Raton, FL 33487-2742

and by CRC Press
2 Park Square, Milton Park, Abingdon, Oxon, OX14 4RN

© 2021 D. H. Stamatis

CRC Press is an imprint of Taylor & Francis Group, LLC

Library of Congress Cataloging-in-Publication Data
Names: Stamatis, D. H., 1947- author.
Title: Automotive audits : principles and practices / D. H. Stamatis.
Description: First edition. | Boca Raton, FL : CRC Press/Taylor & Francis
Group, LLC, 2021. | Series: Practical quality of the future : what it
takes to be best in class (BIC) | Includes bibliographical references
and index.
Identifiers: LCCN 2020041457 (print) | LCCN 2020041458 (ebook) |
ISBN 9780367696597 (hardback) | ISBN 9781003142744 (ebook)
Subjects: LCSH: Automobile industry and trade—Quality control—Standards. |
Automobile industry and trade—Auditing.
Classification: LCC TL278.5 .S829 2021 (print) | LCC TL278.5 (ebook) |
DDC 629.222068/5—dc23
LC record available at https://lccn.loc.gov/2020041457
LC ebook record available at https://lccn.loc.gov/2020041458

ISBN: 978-0-367-69659-7 (hbk)
ISBN: 978-1-003-14274-4 (ebk)

Typeset in Times
by codeMantra

In memory of my colleague and good friend

Anthony Roark

Contents

Preface

Over the last 30 years, I have been giving training seminars and conducting audits all over the country and the world. It has been an experience that has enriched me in many ways while it has also given me an opportunity to learn much about many different organizations, cultures, and the process of auditing. When I train, one of the first questions I ask the participants is: "why do we have a quality management system (QMS)?" The response is all over the map, but most of them reply with the classic responses of: Well, we have to standardize processes, produce quality parts, achieve customer satisfaction, maintain effective and efficient operations, and so on. I have never heard the answer of: to *make a profit*. How unfortunate! We have become afraid to admit that the profit (of course, other things are important as well) is a key driver in any organization including the Not-For Profit organizations. The profit is what makes things happen. Without it, there is no need to have an organization. It seems, particularly in the automotive world, we have demonized that word to the point we are afraid of recognizing that every organization has to make money. What else is the purpose of a quality management system if we cannot make money? In every auditor class I have ever taught, I made it a priority to communicate the importance of what we auditors do and placed an emphasis on the purpose and value of audits for any organization.

It is disheartening to see and hear sad stories about the wastefulness of time in performing audits. It is sometimes discouraging to hear claims – even from management – that we do audits to satisfy the third-party auditors; they are no value to productivity; or they are a total waste; *etc.* I am not one of those individuals who think auditing is a waste of time and do not provide a rewarding experience for all concerned. I do believe and that is why this book is written that audits offer a great opportunity to improve by (a) fixing things that present a *gap* from where you are now and where you have to be and (b) present opportunities for new challenges. A≈well-designed and implemented quality management system **will** provide substantial benefits in both productivity and morale for the employees. That QMS, however, has to be validated, and that is where an audit and the auditor come into the picture.

A successful audit depends on a good auditor who knows (a) the process being audited and (b) has the knowledge of the auditing discipline. From my experience, I have developed five rules that successful auditors must have. They are as follows:

1. Like (enjoy doing) it. If you are not comfortable with asking people to explain what they do, how they do it, and how it is effective, auditing is not for you.
2. Like people. If you are not a people person, one who naturally likes people and can get along with others, auditing is not for you.
3. Believe in the process. If you do not believe in audit process and what it does for the organization, auditing is not for you.

4. Like to learn. If you do not like to learn about how other departments or processes work, auditing is not for you.
5. Care about the organization. If your passion and faith in the company is missing, auditing is not for you.

Therefore, I have tried to present in this book some issues that an auditor needs to know for an effective audit in any industry, but primarily in the automotive industry. Hopefully, I have been successful. Specifically, I have covered the following towards the goal of convincing management that audits are important, and that there is a methodology to conducting an excellent audit. So, each chapter presents the following information:

Chapter 1: Legitimate Concerns about Audits: This chapter addresses that "why" audits are necessary.

Chapter 2: Preassessment Preparation: This chapter addresses what are the prerequisites for a "good" audit.

Chapter 3: Risk Considerations in an Audit: This chapter discusses some of the risks that are possible with and without an audit in any organization.

Chapter 4: Audits: This chapter gives a detail explanation with the essentials of what audits are all about with specificity.

Chapter 5: Mandatory Auditing Items: This chapter addresses the mandatory and non-mandatory documents/records of standards and industry requirements.

Chapter 6: Acronyms: This chapter provides a lengthy list of common acronyms used in the automotive industry.

Chapter 7: Methodologies/Tools That the Auditor Should Be Familiar With: This chapter presents a variety of methodologies/tools that any auditor should be familiar so that s/he can ask questions relevant to the situation.

Chapter 8: Performance beyond Specifications: This chapter presents material that any auditor should be pursuing in order to accomplish the role of a "change agent" towards continual improvement.

Chapter 9: Quick View of Auditing: This chapter gives a practical approach to auditing focusing on the primary drivers (indicators) of quality: quality, delivery, cost, and responsiveness.

Chapter 10: Process Approach to Auditing: This chapter focuses on the layered process audit and some other types of audits as derivatives of the LPA.

Epilogue: It summarizes the main points of the audit.

List of Figures

List of Tables

Acknowledgments

For 40 years, I have been practicing quality including auditing. I have met many wonderful individuals that I have learned and "honed" the practice of auditing and other methodologies and tools throughout my career. It is impossible to mention all of them. However, I cannot write a book about auditing without giving thanks to individuals who guided me and taught me not only quality but also the correct way of auditing. So, let me start with.

Mr. P. Biswas who was kind enough to give me permission to use some of his checklist questions, which were based on Mr. Art Lewis' work as well as many others and websites like advisera.com and many others.

Mr. William Bumgardner, my first supervisor who was my mentor in the quality field and taught me the value of inspection, the power of data and legitimate audits. This was in the late 1960s; however, I have never forgotten the principles that he taught me.

Mr. J. LaPolla who patiently taught me the practice of "unbias" audits and a standardized methodology to investigation.

Dr. R. Munro who introduced me to the methodologies of conducting quality practices at Ford Motor Co. including supplier chain concerns.

Mr. J. Nosal who collaborated with me in many endeavors with supplier and ISO issues and evaluating quality practices.

Mr. S. Mitchell who has spent many hours discussing the issues and concerns about auditing and especially third-party auditing. I have included some of his ideas in this work.

Perhaps one of the most influential persons for this work is Dr. P. Panson who enthusiastically pushed me to write this book but also gave me many insights in the context of the material. In fact, he was the one that suggested strongly to close the book with some kind of an outline emphasizing the supply chain and how the auditor can help. I took his advice and have it as the last chapter.

As always, all my work has been supported by my wife Carla who has been an enthusiastic supporter but also a tough editor for my ideas, style of writing, and quite a few times the order of content. I cannot thank her enough.

Author

D. H. Stamatis is the president of Contemporary Consultants Co., in Southgate, Michigan. He is a specialist in Management Consulting, Organizational Development, and Quality Science. He has taught Project Management, Operations Management, logistics, Mathematical Modeling, Economics, Management, and Statistics for both graduate and undergraduate levels at Central Michigan University, University of Michigan, ANHUI University (Bengbu, China), University of Phoenix, and Florida Institute of Technology. With over 30 years of experience in management, quality training, and consulting, Dr. Stamatis has serviced numerous private sector industries including but not limited to steel, automotive, general manufacturing, tooling, electronics, plastics, food, navy, department of defense, pharmaceutical, chemical, printing, healthcare, and medical device. He is a certified Quality Engineer through the American Society of Quality Control, certified Manufacturing Engineer through the Society of Manufacturing Engineers, certified Master Black Belt through IABLS, Inc., and he is a graduate of BSi's ISO 900 Lead Assessor training program. Dr. Stamatis has written over 70 articles, presented many speeches, and participated in both national and international conferences on quality. He is a contributing author on several books and the sole author of 52 books.

Introduction

Traditionally, the most common motivation of any audit has been and continues to be a process by which verification and validation of relevant internal policies and procedures exist and conform to the applicable standards and requirements that the customer has imposed on a given organization. If validation is positive, everyone is happy. However, if there are discrepancies to what the organization says it does and there is no proof of that, then a non-compliance is issued and it must be taken care of in a reasonable timeframe.

A non-compliance is a weakness that will lead to inferior performance of the management system, and such instances must be identified during the audit. On the other hand, conformance means that the organization is adhering to the requirements set forth in its internal procedures, policies, guidelines, and to external requirements set forth by the specification, its customers, and/or adopted industry practices. Basically, it verifies that we are "doing what we say we do."

But this is not all there is and that is why many organizations find auditing a non-value activity. The fundamental purpose of any audit is to go beyond the common purpose and identify areas of weakness and waste. Opportunities for improvement (OFIs) are the areas where improvements in a process, typically associated with results, are obviously possible. For example, (a) situations where an auditor notes high levels of scrap or rework in a production process and (b) an instance where an auditor notes significant drafts or a high level of heat loss in an area during an EMS audit.

At this point, one may say that there is no requirement to do this. Yes, however, if the auditor is knowledgeable and observant, s/he will use judgment and perspective when generating OFIs. This is why the generation of OFIs is more difficult than conformance verification. It takes experience and some level of familiarity with a process to generate meaningful OFIs.

This is a very valid objective of an audit but one that requires significant perspective and objectiveness. Experienced internal auditors, in particular, are in a very favorable position to identify best practices during audits so that they can be communicated and implemented throughout other facilities of the organization, as appropriate. If properly designed, the internal audit program can serve as an element of an internal benchmarking program in addition to its many other benefits. The extent to which these objectives can be met will depend on the maturity of the management system itself, the skill and training of the internal auditors, and the level of support provided by the management team for the realization of these objectives. The end result is gained value.

Value is ultimately defined in terms of the external customer. Value is tied to the products and services provided by the organization. From the customer's perspective, value is why the organization exists. Value is related to specific products that the company expects specific customers to purchase at a specific price and how the performance and quality of these products can be improved while their costs are steadily reduced due to either new technology or process improvements. Value to the customer translates into loyalty for repurchasing the organization's product or service. Conversely, a "non-value-added" activity (NVA) is any activity that does not provide direct value to the customer and which the customer would not be willing to pay for if the activities were itemized. It is considered, waste.

To be sure, many customers are willing to pay for the effort associated with receiving incoming materials, manufacturing, and packaging and shipping. They are even willing to pay for some level of inspection and testing (depending on the customer and their philosophy regarding building it right the first time). However, they would absolutely not be willing to pay for successive layers of over-inspection, containment, rework, scrap, claims processing, correcting inventories, and so on. That is considered waste and all waste (or NVA activities) adds cost, which reduces profitability and the ability to compete. Wastes are categorized into two types. They are as follows:

- **Type 1 waste** is waste associated with actions that provide no value in the eyes of the customer but which are needed to support current operations as they exist today. They are necessary because of inefficiencies in the way the organization currently operates. Examples include many inspection and testing activities. The actions associated with the ongoing improvement and elimination of type 1 waste can be thought of as continual improvement, or *kaizen*.
- **Type 2 waste** is associated with action that provides no value and is not necessary for current operations. This is pure waste. The actions associated with the initial mapping and elimination of type 2 waste from the value stream can be thought of as *kaikaku* or radical improvement.

So, when one talks about auditing, they must recognize that the most important issue is that auditing is a process in and of itself. ISO 19011:2011 (clause 3.1) states that an audit is a "Systematic, independent and documented process for obtaining audit evidence and evaluating it objectively to determine the extent to which audit criteria are fulfilled." As a consequence, there are distinct steps that must be followed in order to execute a successful audit. The process is shown in Table I.1.

Simply put, a process can be described as an activity that transforms, or converts, inputs into outputs. The process model to auditing was introduced with the release of the ISO 9001:2000 quality standard. The older versions of the standard, and many other current quality models, focus almost exclusively on the transformation step of the process. That is, they provide the requirements that must be met and the controls

TABLE I.1

Auditing as a Process

Input	Process	Output
Information	Evaluation	Findings
Audit Planning		Analysis
		Report(s)

to be used during the transformation, with only minimal consideration of the inputs and only generic requirements for the outputs. Companies implemented the earlier models by developing procedures that laid out the steps needed to properly conduct the activity, unfortunately continue on the "transformation" model of the process itself by grouping of requirements into elements, with each major element focused on a major transformation activity (*e.g.* contract review, design control, product identification). The new approach of "process audit" (see Figure I.1) allows us to recognize the system as a whole and identify inefficiencies much easier.

The process approach focuses on the customer, not internal fiefdoms. Organizations have traditionally used a functional approach to management. In this arrangement, work and communications flow vertically within the department or function. When they reach the top, they are often "thrown over the wall" to the next function or department where the output will be used. Communications and synergy between departments are often poor, and problems at the interfaces between departments can result in poor-quality products and services. The focus is on supporting one's own department, its objectives, and goals, rather than on the customer. In the process approach to manage, the focus is on the customer, resulting in more efficient processes that increase customer satisfaction.

The ISO model emphasizes understanding how the outputs of the process support other processes needed to satisfy the customer. Indeed, these outputs become inputs into these downstream processes and as such also need to be monitored and controlled as inputs. To truly embrace the process concept, the supplier of these outputs would go to these internal "customers" and determine what they need in terms of the quality, information, timeliness, accuracy, format, and other attributes of these deliverables in order to fully meet their needs. This would continue until all of the organization's important processes were aligned to meet both their external and internal customer's needs and expectations. In effect, what is created are *systems of processes* all oriented and operating together towards the achievement of system goals, most typically to satisfy the customer.

As we have already mentioned, most problems occur at the interface between activities or processes. This is why it is important to manage the system as a network of interrelated processes. Focusing on isolated functions does not adequately monitor

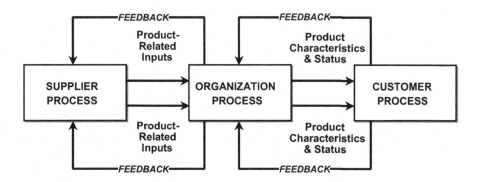

FIGURE I.1   The classic approach to process audit.

the interactions that occur between processes, at the interfaces, which are where problems occur.

In the final analysis, audits help the organization to identify their internal OFIs. OFIs are any areas where improvements in a process, typically associated with results, are obviously possible. That is why it is imperative to have auditors with experience and some level of familiarity with a process to generate meaningful OFIs.

In the same track, audits are the vehicle of recognizing the need for best practices (BP). BPs are very valid objective of an audit but one that requires significant perspective and objectiveness. Experienced internal auditors in particular are in a very favorable position to identify best practices during audits so that they can be communicated and implemented throughout other portions of the organization, as appropriate. If properly designed, the internal audit program can serve as an element of an internal benchmarking program in addition to its many other benefits. On the other hand, properly designed and implemented external (third party) audits can give credence and confidence to the customer of the organization that is being audited as having a good QMS. The extent to which these objectives can be met will depend on the maturity of the management system itself, the skill and training of the internal/external auditors, and the level of support provided by the management team for the realization of these objectives.

1 Legitimate Concerns about Audits

Whether we like it or not, in the last 5 years or so, we see a mistrust towards auditing and their significance of improvement to an organization. The loud noises about "no value added," "waste of time," and other epithets are increasing in both literature and among professionals in seminar settings as well as social events. Why is that happening? Primarily because even though the standards and requirement are increasing at an exponential level, no one is really held accountable for not following them. Quality has taken a back seat to production/profit and mediocrity seems to have become *the new standard* (Smith, 2019, p. 13). It has become a "routine" check with an attitude of "well, this is one more thing we have to do." Many organizations have drop certification according to the ISO, and many more have dropped the pursuit of excellence with the Malcolm Baldrige National Quality Award. An excellent review on this topic is presented by De Carvalho and Sampalo (2020, pp. 43–49).

In the field of quality, as in any other profession, we do indeed have ethics that promote honesty, integrity, safety, and customer satisfaction. To all these categories, we have and continue to generate standards (international, industry, and specific organizational). However, if one looks at the data (statistics) for the last three centuries, we are going to see that as these standards, specifications, and regulations are increasing, not only similar problems are repeated but also are increasing more than ever. For a list of some of the catastrophic examples, see Stamatis (2020). For a more current list, let us see the following examples, specifically in the automotive industry:

> In the auto industry we are witnessing a record number of recalls in 2016 reaching 52,985,779 in total. That's a steep number that was plumped up in large part due to the massive Takata airbag campaign that ultimately drove the Japanese company into bankruptcy. Specifically, Howard (2019) reports that "Ford workers break their silence on faulty transmissions: 'My hands are dirty. I feel horrible.' Ford knew Focus, Fiesta models had flawed transmission, sold them anyway. They knew the truth and kept quiet." In 2017, Wong (2017) reports that the number of recalls was about 28,146,661 with Fiat Chrysler, Honda, Ford, Hyundai and BMW leading the way. That's not a small number, but it's a far cry from the 2016 levels. [It is very interesting that even GE has issues with major quality problems to the tune of $1 billion dollars due to facing potential costs in its jet engine unit from the grounding of Boeing Co's 737 MAX airliner (Scott and Ajmera, 2019)].

Here's a look at some of the biggest and noteworthy auto recalls issued in 2018 as reported by Masterson (2019). In 2018, he reported 1.6 million 2015–2018 Ford F-150; 1.3 million 2014–2018 Ford Fusion, Lincoln MKZ; 1.3 million 2012–2018 Ford Focus; 807,000 2010–2014 Toyota Prius, Prius v; 507,600 2010–2013 Kia Forte, Optima, Hybrid, Sedona; 504,000 2013–2016 Ford Escape, Fusion; 343,000 2012–2017 Audi A4, A4 Allroad, A5, A6, Q5; 240,000 2017–2018 Chrysler Pacifica;

232,000 2018 Honda Accord, 2019 Insight; and 215,000 2015–2018 Nissan and Infiniti vehicles.

In 2019, Masterson (2019a) reported 556,400 2019–2020 Chevrolet Silverado 1500, HDs and GMC Sierra 1500, HDs; 528,600 2011–2013 Dodge Durangos, Jeep Grand Cherokees; 394,000 Nissan Maximas, Muranos, Pathfinders and Infiniti QX60s; 135,700 2019–2020 Ford F-150s; and 72,700 2019 Ford Rangers. For additional information on recalls, see https://www.cars.com/news/recalls/. Retrieved on December 6, 2019.

Johnson (December 19, 2019) and Johnston (2019) reported that General Motors is recalling more than 814,000 pickup trucks and cars in the United States to fix problems with electronic brake controls and battery cables. Furthermore, Isidore (2015) reported that General Motors closed the books on its epic year of recalls, saying they cost the company $4.1 billion in repair costs, victim compensation, and other expenses.

On the other hand, Selby (December 19, 2019) from *Consumer Reports* is more explicit of the recall reporting that the first recall covers nearly 464,000 Cadillac CT6 sedans and Chevrolet Silverado 1500 and GMC Sierra 1500 pickup trucks from 2019. The second recall covers over 350,000 2019 and 2020 Silverado and Sierra 1500 pickups. The problem? A cable connecting the battery and alternator may have too much glue on it, and the second problem is electronic brake controls which can cause stability problems with the brakes. The first problem can interrupt the electrical connection and possibly cause the trucks to stall or even catch fire. The second disables the anti-lock brake system causing the instability (Selby (2019): https://www.myarklamiss.com/news/consumer-reports/gm-recalls-814k-pickups-cars-to-fix-brake-battery-problems/. Retrieved on December 20, 2019).

To make things worse, Krisher (2020) reports that Ford has issued a recall involving 2.5 million vehicles including Focus and Fusion for door latch problem and brake fluid leaks. What is incredible about this recall is that Ford has had nagging quality troubles with the latches, some car transmissions, and other issues that have hurt its bottom line.

The company said Wednesday (August 12, 2020) the previous door latch recalls were done because of defective pawl spring tabs that could crack and fail in high temperatures. Usually the doors won't close if there's a failure, but if they do close, they could open again while the vehicles are in motion. The previously recalled vehicles may not have had the latches replaced, or repairs may not have been done correctly, the company said in a statement.

Obviously, these failures are system failures, and as a consequence, both of these problems should have been caught by a thorough internal audit and certainly a third-party audit and also, had a thorough design review, reliability analysis, design failure mode and effect analysis (DFMEA), advanced product quality planning (APQP), or some kind of simulation modeling. It is hard to believe that no one knew of this before launching the vehicles. One wonders with so many standards and regulations, why do we have so many problems? There are seven options:

- By far, the first and most critical and common is the manipulation of data. Wrong strategy, wrong definition, wrong selection, wrong methodology, inadequate analysis, and bias presentations.

- The second is that management knows about it, but as the workers reported in Howard (2019), "They knew the truth and kept quiet."
- The third is that the significant tools, *i.e.* APQP, FMEA (failure mode and effect analysis), and appropriate (applicable) problem-solving techniques are not conducted, or if they are, they are not done correctly. They are done to have a checklist completed for "things done" and not for improvement.
- The fourth is (perhaps the predominant one) *production is priority* and nothing else matters. *Quality and even safety are secondary considerations*, although they are preached as priority. They are evaluated after the fact, rather than being evaluated up front.
- The fifth reason why things are becoming problems for the customer I believe is the *"laissez-faire"* attitude of organizations towards audits. The intent of "audits" is to identify weakness in a system. However, in most organizations – if not all of them – audits are taken place because some "standard" or "regulation" or "specification" calls for them. Unfortunately, as many millions of dollars are spent on audits, the benefits are questionable at best, given the problems and recalls of many products. This is verified by the amount of rejects, recalls, and issue notifications to customers. Stoop (2020) reports that according to the National Safety Council, workplace fatalities have risen 17% since 2009 after decades of steady improvement in occupational safety, outpacing workforce growth over that period. At the same time, international, industrial, and customer-specific standards/ requirements have increased in both volume and complexity.
- The sixth reason is that the focus of management is *short-term gains* based on *"quarterly earnings"* and not long-term, real improvements and productivity and earnings.
- The seventh reason is that the audits – at all levels – were not effective for whatever reason.

In addition to these seven reasons, it is worth examining the multiple variables associated with safety recalls the National Highway Traffic Safety Administration (NHTSA, 2018) has identified as potential risks. All have to do with "some kind" of failure that *could have been caught, but it did not*. The variables according to NHTSA are as follows:

1. The manufacturer;
2. The age of the oldest affected vehicle;
3. The vehicle type involved (*i.e.* passenger cars, lights trucks, Multi-Purpose vehicles (MPVs));
4. The component category;
5. The recall safety risk description includes the word "crash";
6. The recall safety risk description includes the word "fire";
7. The recall safety risk description includes the word "death";
8. The recall safety risk description includes the word "injury";
9. The recall safety risk description includes the word "serious";
10. The year the recall was initiated; and
11. The number of vehicles affected by the recall.

So, one can see that indeed the old adage has some truth to it. That is, figures do not lie, but liars figure. It is imperative therefore to at least mention the role of the "data handlers," the people who handle data in our education or industrial system are expected to do many things – and do them all well. Most of these individuals are trained well and provide the appropriate and applicable studies with statistically sound results to their respective organizations. Other data handlers have non-instructional leadership or administrative support roles, and they are short on both knowledge and practical experience of statistics. Still others provide highly skilled technical or data expertise that contributes to the effective and efficient operation of their enterprise. Regardless of an individual's job title, working in a modern organizational environment demands (a) unwavering adherence to codes of appropriate conduct, operating expectations, and professional standards and (b) some level of statistical knowledge.

A "data handler" is defined here as anyone involved or has excess in the usage of data (from data definition, selection of the methodology, analysis of the results, and reporting of the results) in any organization. Honest data handlers can be trusted to maintain objectivity and uphold an organization's data procedures and protocols even when it requires extra effort, is not convenient, or otherwise runs counter to their own **personal interests**. Two horrendous examples of mishandling data (tests) recently are as follows:

1. The combination of tests to identify "a" formulating policy by Center for Disease Control (CDC) regarding the corona virus. The Centers for Disease Control and Prevention (CDC&P) acknowledged on Thursday that it is combining the results from viral and antibody COVID-19 tests when reporting the country's testing totals, despite marked differences between the tests. First reported by NPR's WLRN station in Miami, the practice has drawn ire from the US health experts who say combining the tests inhibits the agency's ability to discern the country's actual testing capacity. "You've got to be kidding me," Ashish Jha, director of the Harvard Global Health Institute, told *The Atlantic*. "How could the CDC make that mistake? This is a mess" (Johnson, 2020). Madrigal and Meyer (May 21, 2020).

2. *Lancet* Study (AP, 2020). One of the most prestigious health journals published a hydroxychloroquine study, even though they knew there were serious concerns about the data. The AP reported the study thusly: Concerns are mounting about studies in two influential medical journals on drugs used in people with corona virus, including one that led multiple countries to stop testing a malaria pill.

   a. *The New England Journal of Medicine* issued an "expression of concern" on Tuesday on a study it published on May 1 (2020) that suggested widely used blood pressure medicines were not raising the risk of death for people with COVID-19. The study relied on a database with health records from hundreds of hospitals around the world. "Substantive concerns" have been raised about the quality of the information, and the journal has asked the authors to provide evidence it's reliable, the editors wrote.

b. The same database by the Chicago company Surgisphere Corp. was used in an observational study of nearly 100,000 patients published in *Lancet* that tied the malaria drugs hydroxychloroquine and chloroquine to a higher risk of death in hospitalized patients with the virus. *Lancet* issued a similar expression of concern about its study on Tuesday, saying it was aware "important scientific questions" had been raised. For more information on the faulty data, see Ramsey (2020).

So, regardless of a data handler's role in an organization, consistently and continuously demonstrating honesty, integrity, and professionalism are of paramount importance. These qualities, more than any other characteristic or trait, serve as the foundation of ethical behavior, not only in quality engineering but also in all disciplines and in all inquiries of furthering knowledge. Hopefully, internal and external audits will identify system gaps and "fix" them, but also as a result of these findings, the organization will develop "preventive" measures to AVOID future issues, concerns, and even problems.

2 Preassessment Preparation

I remember my Boy Scout motto of *Be Ready*. It has served me well over the years; however, I have added the words *for the unexpected*. So now the Boy Scout motto has become *Be Ready for the Unexpected*. For any type of audit, preparation is the foundation of excellence. The more prepared one is, the more successful the audit will be. Therefore, in this chapter, we present an example of generic questions to serve as a guide.

To be sure, one can never be 100% prepared for all combinations and expectations of a process. So, it is a good practice to have a list (a map, of sorts) that will guide the auditor to find (through probing) possible gaps in both expected and unexpected situations.

A TYPICAL INTERNAL PREASSESSMENT SURVEY

The following questions are intended to be used only as guideline in a given organization. The questions are designed to identify any shortcomings in your system and to allow you to plan accordingly. They are not meant to be used as a formal checklist for any organization, since the official checklist is prepared by the auditors themselves and/or representatives of the Registrar. The list is based on Stamatis (1996) and Grossman (1995, pp. 34–35).

1. Does your company have a written quality policy that describes management's commitment to quality and objective for achieving quality in every part of the company's operation?
2. Has your management group endorsed the quality policy and communicated the policy to all employees?
3. Is there an approved organization chart showing who is responsible for all work that affects the quality of the product or service that your company produces?
4. Are the functions and job specifications for personnel who affect the quality of the product or service clearly defined?
5. Are the technical and personnel resources that are needed for the inspection, testing, and monitoring of the production of the product or service made available by management?
6. Are the technical and personnel resources that are needed for the inspection, testing, and monitoring of the product or service during its life cycle made available by management?
7. Are periodic audits of the quality system completed as often as necessary to keep each part of the system in control? (Internal audits – processed or layered audits)

8. Are periodic audits of the manufacturing processes completed as often as necessary to keep each process in control?

9. Are periodic audits of the product or service that your company produces completed as often as necessary to ensure that the quality of the product meets customer requirements?

10. Are the results of the audit communicated to management and to those employees who affect quality?

11. Has your company appointed a coordinator to be responsible for monitoring the quality system and calling attention to the deficiencies?

12. Are quality reviews held at appropriate intervals?

13. Are the results of the audits recorded and maintained?

14. Are procedures written for each activity that affects quality? Are they appropriately maintained? Are they easily accessed by the employees?

15. Does your company have a plan for achieving and maintaining quality?

16. Does your company audit and evaluate its progress in achieving the objectives listed in the quality plan?

17. Are customer needs identified and communicated to all employees who affect the quality of the product?

18. Do employees know what they have to do on the job to provide the desired level of quality in the product or service?

19. Are the customer requirements for product and service quality adequately defined in the contract with the customer?

20. Are customer contracts reviewed for accuracy?

21. Are records of the customer reviews maintained?

22. Are incomplete and ambiguous requirements resolved before design or production?

23. Are all applicable and appropriate documents reviewed before they are released for use?

24. Do you have an obsolescent policy? Do you follow it? What is your policy for discarding it?

25. Do your procedures and instructions describe what is actually done on the job, now?

26. Do you have document control? Do you follow it?

27. Do you have a certification program for your suppliers? If not, how do you approve your suppliers?

28. Do you keep performance records from your suppliers? Do you perform regular analysis with the data? Do you communicate the information to your supplier base?

29. Does someone check all incoming supplies and equipment to verify that you have indeed received the correct resources to do the job and that they meet the defined requirements?

30. Do you maintain the list of approved suppliers?

31. Do you audit your suppliers?

32. Do you use systematic methods to identify and plan production processes and (if appropriate) equipment and product installation processes?

33. Do your employees use their own tools? How do you make sure they are calibrated?
34. Do you do calibration?
35. Do you have written setup and process instructions?
36. Do you have preventive maintenance?
37. Do you have written standards for workmanship and criteria for meeting the standards?
38. Do your employees follow job procedures and instructions?
39. Do your employees follow unwritten procedures or instructions?
40. Do the procedures and instructions describe the way employees do their jobs now?
41. Do you record tooling repairs to ensure process control?
42. Do you have written procedures to ensure that incoming products are not used or handled before an inspection or other form of verification proves that these products meet specified requirements?
43. Are inspection procedures carried out in accordance with written instructions and your company's quality plan?
44. Do you have written procedures to identify incoming material that may have been released before it was inspected because of urgent production purposes?
45. Do you maintain a receiving inspection history or log?
46. Does your company collect and maintain records to prove that you have met customer requirements?
47. Do you have written instructions for inspecting and testing?
48. Are nonconforming products identified and separated so that they are not sent to customers? What is your quarantine policy?
49. Are there written procedures to verify that all final inspections and tests are completed before products are sent to customers?
50. Are there written procedures for calibration and maintenance of inspection, measurement, and test equipment that show calibration frequency?
51. Do you have a system to identify the inspection or test status of products during manufacturing?
52. Is there a documented procedure for identifying and separating rejected material to prevent inadvertent use of nonconforming products?
53. Is there a method of recording the rejected material and the disposition of such material? Are there documents to support that the method is being followed?
54. Is there a method for requesting a deviation from the customer? Is it being followed? Is there documentation to support the practice?
55. When a waiver of change or a deviation has been authorized by the customer, is that information recorded and maintained?
56. Is there an analysis of nonconformities?
57. Are there procedures for ensuring that effective corrective actions are carried out?
58. Are there procedures from preventing damage to products as they are handled?

59. Are in-stock products inspected at periodic intervals?
60. Is there a written procedure for identifying, collecting, indexing, filing, maintaining, and disposing of quality-related records?
61. Are quality records maintained so that the achievement of their required levels of quality can be demonstrated to customers and to your management team?
62. Are quality records stored in an accessible place? Are they retrievable?
63. Are quality records accessible to your customers for their review?
64. Do you have a retention policy? Is it written? Is it being followed?
65. Are quality audits performed as defined in your procedures?
66. Do the appropriate personnel take timely corrective actions? Are their actions recorded?
67. Do training and development plans exist for all employees who have an impact on the organization, product, or service?
68. Are records maintained to show who attended training, when they attended, and their success in learning the skills?
69. Are there written procedures and instructions for follow-up service? Does appropriate maintenance exist for these procedures? Do they meet the requirements?
70. Is there a method of establishing the need for statistical techniques? How do you maintain control in your processes?

The reader of this checklist should notice that the accepted answer for all questions is a "YES" response. This is a very shallow response and in fact in most cases a questionable response. Therefore, this list is given here to be used ONLY as the "spring board" for other follow-up deep-dive questions. The intent is to help the auditor "break the ice" and to help the operator be at ease. In essence, it gives confidence to both the auditee and auditor of a positive experience. It avoids the "got you" moment.

Furthermore, I do not claim to be the original author to many of the questions of the presented checklists. I would like to thank all the original writers. However, I do not know all of them, since most of them have been selected from several lists with and without sources. Many thanks, however, go to Mr. P. Biswas who was kind enough to give me permission to use some of his questions which were based on Mr. Art Lewis' work as well as many others and websites like advisera.com and many others.

3 Risk Consideration in Audits

Risk is something that comes to all of us, and it is always good to account for anything that we do. Let us remember Murphy's law which exemplifies why we should be concerned with risk. Murphy reminds us *that if anything can go wrong, it will.* In the world of quality, many things can go wrong in both expected and unexpected situations. That is why we find in all regulations, standards, requirements, and specifications the concept of risk.

In the most elementary level when we address risk in any audit endeavor, we mean that a thorough examination and evaluation that an organization has as part of their system, to avoid safety and any hazardous conditions. If they do, there must be a system to correct and prevent such deficiencies.

In an audit environment, it is the responsibility of the auditor (whether internal or external) to focus on the effectiveness of the risk prevention methods and evaluate management's responsibility and commitment throughout the organization. All standards and industrial requirements – to some extent – require a plan in place to assess how risks are identified and what needs to be done to either minimize or eliminate them.

Perhaps one of the major mistakes that many organizations make is the notion that risk audits are a one-time event. That is completely false. Risk audits are an ongoing activity to ensure that projects are on track and risk policies are being followed. However, in the course of an audit, the particular issues of the risk should be addressed following a general outline. The outline is a composite document generated by the auditor, employees, and management – as appropriate and applicable: The outline should address/cover at least the following:

1. All possible hazards and past safety issues, if corrected, were they effective?
2. Similar past issues, concerns, and problems how they are handled. Be especially aware of how the employees and management team react and behave to risk (safety and hazard events).
3. Current policies and procedures that are up to date and cover all standards, regulations, and corporate mandates.
4. Appropriate and applicable training has taken place and it is being followed.

Specifically, there are six steps to conduct a focused risk audit. They are as follows:

1. *Decide the risk team*: Who are the team members to conduct the audit? Do they have appropriate and applicable knowledge about the risk and the process they are going to audit?
2. *Interviewing team members*: Once the risk team requirements have been defined, now it is time to make sure that the individuals have the qualifications

to perform the audit. It is imperative that they know all the stakeholders and the ramifications of safety and hazards upon them. It is very important NOT to assume that all the selected team members are aware of the procedures and know what to do in an emergency.

3. *Critical success factors*: This is a key component of the audit because unless the success factors have been identified, there is no way to define success. There is no generic success factor list as they depend on the individual process, organization, and requirements.

4. *Gathering evidence*: The collection of evidence is through: (a) scheduled interviews with both employees and managers of the process and other stakeholders as appropriate and applicable; (b) reviewing standards, regulations, and requirements; and (c) reviewing policy, procedures, instructions, and compare them to the answers from the interviews.

5. *Analyzing evidence and creating a report*: The analysis demands that the auditor evaluates the evidence for timeliness, goals, objectives, and gaps. As an auditor, you may make recommendations to improve the risks (for both safety and hazards) of the process.

6. *Follow-up audits*: Once the audit is complete, you may want to conduct a follow-up audit to make sure that all gaps and recommendations have been "fixed." The follow-up audits need to be conducted with the same vigor as the original audit. Of course, these follow-up audits are in additional to the regularly scheduled audits during the year.

RISK

Risk management principles are effectively utilized in many areas of business, quality, and government including manufacturing, finance, insurance, occupational safety, public health, pharmaceutical, pharmacovigilance, and by agencies regulating these industries. In general terms, risk is defined as the combination of the probability of occurrence of harm and the severity of that harm. However, achieving a shared understanding of the application of risk management among diverse stakeholders is difficult because each stakeholder might perceive different potential harms, place a different probability on each harm occurring, and attribute different severities to each harm. So, even though it is difficult to quantify risk in its many forms, it is imperative that we make every effort to identify it and measure it as much as possible. To do this evaluation, we at least have two principles that must be followed:

The evaluation of the risk to quality should be based on scientific knowledge, and

The level of effort, formality, and documentation of the quality risk management process should be commensurate with the level of risk.

On the other hand, when we discuss *risk* in terms of quality, we have in mind a wider spectrum of what risk is all about and that is why practically all international, industrial as well as specific customer requirements are identified and required to be addressed.

So, for a full understanding of what *risk* is in quality terms, we must, above all, recognize that quality risk is unquestionably a management process. That is: in no uncertain terms, it is a systematic process for the recognition, definition, assessment, control, communication, and review of risks to the quality of product across the product life cycle. Of course, these functions are managerial functions and as such depend on the management's commitment and actions to be taken seriously. There are many models in literature that one can follow. A simple one (which of course may be modified to reflect individual organizational needs) is the following:

- Initiate the risk assessment
  - Risk identification
  - Risk analysis
  - Risk evaluation
- Risk control
  - Risk elimination
  - Risk reduction
  - Risk acceptance
- Output of the assessment
  - Risk review
  - Review events
- Take action on specific tasks.

The emphasis on each component of the framework might differ from case to case but a robust process will incorporate consideration of all the elements at a level of detail that is commensurate with the specific risk. It is important here to recognize that the "best results" will be when communication is open to all concerned and the appropriate and applicable tools are used for evaluating the process.

RISK IDENTIFICATION

Risk identification is a systematic use of information to identify hazards referring to the risk question or problem description. Information can include historical data, theoretical analysis, informed opinions (this is very important – especially if it comes from individuals close to the source), and the concerns of stakeholders. A stakeholder may be a government, community, customer, employees, or investor. Risk identification addresses the "What might go wrong?" question, including identifying the possible consequences. Addressing the identification of a particular risk may lead the management and the auditor in more detailed analysis and/or completely different trails.

RISK ANALYSIS

Risk analysis is the estimation of the risk associated with the identified hazards. It is the qualitative or quantitative process of linking the likelihood of occurrence and severity of harms. In some risk management tools, the ability to detect the harm (detectability) also factors in the estimation of risk. In this stage, it is imperative to do a risk versus benefit analysis (*i.e.* PEST, SWOT) to understand the consequences of what the risk taken may result in.

RISK EVALUATION

Risk evaluation compares the identified and analyzed risk against given risk criteria. Risk evaluations consider the strength of evidence for all three of the fundamental questions. In doing an effective risk assessment, the robustness (impervious to noise) of the data set is important because it determines the quality of the output. Revealing assumptions and reasonable sources of uncertainty will enhance confidence in this output and/or help identify its limitations. (An excellent tool for this analysis is the P-diagram.) Uncertainty is due to combination of incomplete knowledge (theoretical and/or practical) about a process and its expected or unexpected variability. Typical sources of uncertainty include gaps in knowledge, gaps in process understanding, sources of harm (*e.g.* failure modes of a process, sources of variability), and probability of detection of problems.

The output of a risk evaluation may be either a quantitative estimate of risk or a qualitative description of a range of risk. When risk is expressed quantitatively, a numerical probability is used. Alternatively, risk can be expressed using qualitative descriptors, such as "high," "medium," or "low," which should be defined in as much detail as possible. Sometimes a risk score is used to further define descriptors in risk ranking. In quantitative risk assessments, a risk estimate provides the likelihood of a specific consequence, given a set of risk-generating circumstances. Thus, quantitative risk estimation is useful for one particular consequence at a time. Alternatively, some risk management tools use a relative risk measure to combine multiple levels of severity and probability into an overall estimate of relative risk. The intermediate steps within a scoring process can sometimes employ quantitative risk estimation.

RISK CONTROL

Risk control includes decision-making to reduce and/or accept risks. The purpose of risk control is to reduce the risk to an acceptable level (low probability). (A good tool for this is the *decision tree* methodology which is based on conditional probability of events.) The amount of effort used for risk control should be proportional to the significance of the risk. Decision makers might use different processes, including benefit–cost analysis, for understanding the optimal level of risk control. (Here it is very important that one must recognize that there are some situations that *the risk will never be eliminated*. This means that the analysts of the specific risk must focus on *reduction and or control of the risk* given "some doable" parameters). Risk control might focus on the following questions:

- Is the risk above an acceptable level?
- What can be done to reduce or eliminate risks?
- What is the appropriate balance among benefits, risks, and resources?
- Are new risks introduced as a result of the identified risks being controlled?

Risk reduction focuses on processes for mitigation or avoidance of quality risk when it exceeds a specified (acceptable) level. Risk reduction might include actions taken to mitigate the severity and probability of harm. Processes that improve the detectability of hazards and quality risks might also be used as part of a risk control strategy. The implementation of risk reduction measures can introduce new risks into the system or increase the significance of other existing risks. Therefore, it might be appropriate to revisit the risk assessment to identify and evaluate any possible change in risk after implementing a risk reduction process.

No matter how you look at the risk acceptance, in the end it always remains a decision to accept risk – sometimes including fatalities. Risk acceptance can be a formal decision to accept the residual risk or it can be a passive decision in which residual risks are not specified. For some types of harms, even the best quality risk management practices might not entirely eliminate risk. In these circumstances, it might be agreed that an appropriate quality risk management strategy has been applied and that quality risk is reduced to a specified (acceptable) level. This (specified) acceptable level will depend on many parameters and should be decided on a case-by-case basis. Quite often, these parameters are defined by governmental laws and regulations.

RISK COMMUNICATION

Risk communication is the sharing of information about risk and risk management between the decision makers and others. Parties can communicate at any stage of the risk management process. The output/result of the quality risk management process should be appropriately communicated and documented. Communications might include those among interested parties (*e.g.* regulators (national, state, or local), industry, within a company, community, employees, and customers). The included information might relate to the existence, nature, form, probability, severity, acceptability, control, treatment, detectability, or other aspects of risks to quality and must be documented. The management of any organization dealing with risk(s) must understand that any communication and management decision(s) dealing with risk(s) may affect other decisions within the organization depending on the ultimate resolution dealing with *a* specific risk. To be sure, a single resolution of a particular risk may have a cumulative effect on other decisions based on specific regulations.

RISK REVIEW

Risk management should be an ongoing part of the quality management process. A mechanism to review or monitor events should be implemented. The output/results of the risk management process should be reviewed to take into account new knowledge and experience. Once a quality risk management process has been initiated, that process should continue to be utilized for events that might impact the original quality risk management decision, whether these events are planned (*e.g.* results of product review, inspections, audits, change control) or unplanned (*e.g.* root cause from failure investigations, recall). The frequency of any review should be based upon the level of risk. Risk review might include reconsideration of risk acceptance decisions. (For more detailed information, see Hall, 1998; Stamatis, 2014, 2015c, 2019.)

METHODS AND TOOLS USED IN RISK MANAGEMENT

Quality risk management supports a scientific and practical approach to decision-making. It provides documented, transparent, and reproducible methods to accomplish steps of the quality risk management process based on current knowledge about assessing the probability, severity, and, sometimes, detectability of the risk. Traditionally, risks to quality have been assessed and managed in a variety of informal ways (empirical and/or internal procedures) based on, for example, compilation of observations, trends, and other information. Such approaches continue to provide useful information that might support topics such as handling of complaints, quality defects, deviations, and allocation of resources. An organization can assess and manage risk using recognized risk management tools and/or internal procedures (*e.g.* standard operating procedures). The list of tools available for organizations as well as auditors is too long. However, some of the key tools and methodologies are listed here.

BASIC RISK MANAGEMENT FACILITATION METHODS

Some of the simple and popular techniques that are commonly used to structure risk management by organizing data and facilitating decision-making are as follows:

- Brainstorming
- 3×5 WHYs
- Flowcharts
- Check sheets
- Process mapping
- Value stream mapping
- Cause and effect diagrams (also called an Ishikawa diagram or fish bone diagram).
- All of them may be found in many statistical process control books and/or as individual books or articles in the general literature domain.

FAILURE MODE EFFECTS ANALYSIS (FMEA)

FMEA provides for an evaluation of potential failure modes for designs and processes and their likely effect on outcomes and/or product performance. Once failure modes are established, risk reduction can be used to eliminate, contain, reduce, or control the potential failures. FMEA relies on product and process understanding. FMEA methodically breaks down the analysis of complex processes into manageable steps. It is a powerful tool for summarizing the important modes of failure, factors causing these failures, and the likely effects of these failures. FMEA can be used to prioritize risks and monitor the effectiveness of risk control activities. FMEA can be applied to equipment and facilities and might be used to analyze a manufacturing operation and its effect on product or process. It identifies elements/operations within the system that render it vulnerable. The output/ results of FMEA can be used as a basis for design or further analysis or to guide resource deployment. For a detailed discussion on FMEA, see Stamatis (2003, 2015a). The reader and the auditor should be very careful here. The traditional approach has been to follow a flow of failure – effect and then the root cause with an assigned Risk Priority Number (RPN) for each root cause of the failure. The new approach of the Automotive Industry Action Group (AIAG)/ Verband der Automobilindustri (VDA) is to identify the effect, follow it with the failure, and then the root cause. The evaluation is also more subjective and much more detailed than the traditional one.

FAILURE MODE, EFFECTS, AND CRITICALITY ANALYSIS (FMECA)

FMEA might be extended to incorporate an investigation of the degree of severity of the consequences, their respective probabilities of occurrence, and their detectability, thereby becoming a Failure Mode, Effects, and Criticality Analysis (FMECA). In order for such an analysis to be performed, the product or process specifications should be established. FMECA can identify places where additional preventive actions might be appropriate to minimize risks. (It is an excellent tool for environmental and OH&S situations (for more detailed info, see Stamatis, 2014).)

FMECA application should mostly be utilized for failures and risks associated with manufacturing processes; however, it is not limited to this application. The output of an FMECA is a relative risk "score" for each failure mode, which is used to rank the modes on a relative risk basis.

FAULT TREE ANALYSIS (FTA)

Whereas the FMEA methodology is a bottom-up approach, the FTA is a top-bottom approach. It is more holistic and it provides more specificity for the failures as they apply to components, subsystems, and or systems. The FTA tool is an approach that assumes failure of the functionality of a product or process. This tool evaluates system (or subsystem) failures one at a time but can combine multiple causes of failure by identifying causal chains. The results are represented pictorially in the form of a tree of fault modes. At each level in the tree, combinations of fault modes are described with logical operators (AND, OR, *etc.*). FTA relies on the experts' process understanding to identify causal factors.

FTA can be used to establish the pathway to the root cause of the failure. FTA can be used to investigate complaints or deviations in order to fully understand their root cause and to ensure that intended improvements will fully resolve the issue and not lead to other issues (*i.e.* solve one problem yet cause a different problem). FTA is an effective tool for evaluating how multiple factors affect a given issue. The output of an FTA includes a visual representation of failure modes. It is useful both for risk assessment and in developing monitoring programs. For more information, see Stamatis (2010).

HAZARD ANALYSIS AND CRITICAL CONTROL POINTS (HACCP)

HACCP is a systematic, proactive, and preventive tool for assuring product quality, reliability, and safety. It is a structured approach that applies technical and scientific principles to analyze, evaluate, prevent, and control the risk or adverse consequence(s) of hazard(s) due to the design, development, production, and use of products.

HACCP consists of the following seven steps:

a. Conduct a hazard analysis and identify preventive measures for each step of the process
b. Determine the critical control points
c. Establish critical limits
d. Establish a system to monitor the critical control points
e. Establish the corrective action to be taken when monitoring indicates that the critical control points are not in a state of control
f. Establish system to verify that the HACCP system is working effectively
g. Establish a record-keeping system.

HACCP might be used to identify and manage risks associated with physical, chemical, and biological hazards (including microbiological contamination). HACCP is most useful when product and process understanding is sufficiently comprehensive to support identification of critical control points. The output of a HACCP analysis is risk management information that facilitates monitoring of critical points not only in the manufacturing process but also in other life cycle phases. For a detailed discussion, see Stamatis (2014).

HAZARD OPERABILITY ANALYSIS (HAZOP)

HAZOP is based on a theory that assumes that risk events are caused by deviations from the design or operating intentions. It is a systematic brainstorming technique for identifying hazards using so-called guide words. Guide words (*e.g.* No, More, Other Than, Part of) are applied to relevant parameters (*e.g.* contamination, temperature) to help identify potential deviations from normal use or design intentions. HAZOP often uses a team of people with expertise covering the design of the process or product and its application.

HAZOP can be applied to manufacturing processes, including outsourced production and formulation as well as the upstream suppliers, equipment, and facilities for drug substances and drug products. It has also been used primarily in the pharmaceutical industry for evaluating process safety hazards. As is the case with HACCP, the output of a HAZOP analysis is a list of critical operations for risk management. This facilitates regular monitoring of critical points in the manufacturing process. For more detailed information, see Stamatis (2014).

PRELIMINARY HAZARD ANALYSIS (PHA)

PHA is a tool of analysis based on applying prior experience or knowledge of a hazard or failure to identify future hazards, hazardous situations, and events that might cause harm, as well as to estimate their probability of occurrence for a given activity, facility, product, or system. The tool consists of

a. The identification of the possibilities that the risk event happens,
b. The qualitative evaluation of the extent of possible injury or damage to health that could result,

c. The relative ranking of the hazard using a combination of severity and likelihood of occurrence, and

d. The identification of possible remedial measures.

PHA might be useful when analyzing existing systems or prioritizing hazards where circumstances prevent a more extensive technique from being used. It can be used for product, process, and facility design as well as to evaluate the types of hazards for the general product type, then the product class, and finally the specific product. PHA is most commonly used early in the development of a project when there is little information on design details or operating procedures; thus, it will often be a precursor to further studies. Typically, hazards identified in the PHA are further assessed with other risk management tools such as those in this section. For more detailed information, see Stamatis (2014). For a very good example to follow for a PHA see: Preliminary hazards analysis for *The Linac Coherent Light Source Ultrafast Science Instruments (LUSI) Project*. Stanford Linear Accelerator Center for the Department of Energy Document No. PM-391-000-99. Revision 2, February 29, 2008.

RISK RANKING AND FILTERING

Risk ranking and filtering is a tool for comparing and ranking risks. Risk ranking of complex systems typically involves evaluation of multiple diverse quantitative and qualitative factors for each risk. The tool involves breaking down a basic risk question into as many components as needed to capture factors involved in the risk. These factors are combined into a single relative risk score that can then be used for ranking risks. "Filters," in the form of weighting factors or cut-offs for risk scores, can be used to scale or fit the risk ranking to management or policy objectives.

Risk ranking and filtering can be used to prioritize manufacturing sites for inspection and/or audit by regulators or industry. Risk ranking methods are particularly helpful in situations in which the portfolio of risks and the underlying consequences to be managed are diverse and difficult to compare using a single tool. Risk ranking is useful for management to evaluate both quantitatively-assessed and qualitatively-assessed risks within the same organizational framework. For more information, see Stamatis (2014).

SUPPORTING STATISTICAL TOOLS

Statistical tools can support and facilitate quality risk management. They can enable effective data assessment, aid in determining the significance of the data set(s), and facilitate more reliable decision-making. A listing of some of the principal statistical tools commonly used is as follows:

- Control charts, for example: acceptance control charts, control charts with arithmetic average and warning limits, cumulative sum charts, Shewhart control charts, weighted moving average, *etc.*
- Design of experiments (DOE)
- Histograms

- Pareto charts
- Process capability analysis
- Six Sigma
- And many more.

There are many sources that anyone may find the above tools. However, one may consider Stamatis (2003) and/or many other sources in books, articles as well as websites.

Why did we address risk in an auditing book? Because the ISO 9001:2015 and IATF 16949:2016 call for it in clauses 4.4.1.2; and 6.1–6.3; and ISO 9004:2018 calls for it in clause 9.5.

4 Audits

OVERVIEW

ISO 8402 defines a quality audit as "a systematic and independent examination to determine whether quality activities and related results comply with planned arrangements and whether these arrangements are implemented effectively and are suitable to achieve objectives" (ANSI/ISO/A8402, 1994). Thus, an audit is a human evaluation process to determine the degree of adherence to prescribed norms and results in a judgment. The norms, of course, are always predefined in terms of criteria (customer specific, industry specific), standards (international standards, industry standards, government regulation), or a combination of them.

To avoid confusion and misunderstandings, it must be understood from the very beginning of the audit that the norms are defined by the management of the organization and the auditor has nothing to do with evaluating the stability and suitability of such norms. However, it is the responsibility of the auditors to evaluate the compliance or the conformance of the organization with those norms. For this reason, the audit is usually performed as a pass/fail evaluation rather than a point system evaluation. (Here we must recognize that some customer-specific requirements do require a point system evaluation such as the Ford Q1 3rd edition.)

As defined then, an audit is an information gathering activity with the purpose of identifying non-compliances or non-conformances (NCs) in the quality management system (QMS) so that improvement(s), corrective action(s), or both may be evaluated and implemented. It is also of paramount importance that prevention actions should be identified and evaluated appropriately. Furthermore, it is essential for any auditor or lead auditor NOT to have preconceived NCs or non-compliances in either volume or significance. The practice of *minor* or *major* should be avoided as both need to be recognized as *gaps* and, therefore, need to be fixed. Above all, at all times, it must be remembered that ALL audits are not public hangings nor instruments for people humiliation. Rather, they are ways to continually optimize the system of a given organization. This is true for the first, as well as for subsequent audits.

Quality audits are neither inspection tools nor verification/validation tools for actual acceptance or rejection of the product or service. The orientation of the audit is *fact finding*; its focus is the *evaluation of the system or process*. The focus of an audit is always prevention and planning, as opposed to inspections where the focus is appraising product quality after the fact. Therefore, quality audits may be internal or external and take one of three forms:

1. *First-party audit*: This audit is conducted by an organization on itself and may be done on the entire organization or part of the organization. It is usually called an *internal audit*.

2. *Second-party audit*: This audit is conducted by one organization on another. It is usually an audit on a supplier by a customer, and it is considered an *external audit*.
3. *Third-party audit*: This audit is conducted by an independent organization (the third party) on a supplier. It can be conducted at the request of a customer or on the initiative of a supplier to gain certification. It is always an *external audit*.

A comparison of the three audits is shown in Table 4.1.

TABLE 4.1

Comparison of Audit Processes

Item	Internal Audit	Second-Party Audit	Third-Party Audit
Definition	Auditing own organization using a planned schedule and trained auditors. May be the result of customer feedback, problem areas, or management requisition	Audit initiated by the client, the organization, or auditee. May be conducted as part of a supplier assurance program	Audit by independent or regulatory body. For certification to the ISO and the industrial standards, it is a mandatory requirement to have an independent audit
Person responsible for planning	Person managing internal audit program. May be the quality manager or the ISO, VDA, or the IATF coordinator	Person managing corporate audit program and/or supplier quality. May be the quality manager or the purchasing manager	Lead auditor assigned to the specific audit
Checklist	Quality manual (optional for the ISO 9001:2015, but highly recommended); procedures, work instructions; customer feedback; previous audit (internal and external) results; and actions taken for closure	Performance records, contract order specific and agreed quality plan audits	Detailed audit plan
Opening meeting	Very informal	Informal to very formal depending on the relationship	Very formal
The AUDIT	Interviews, observations, reviews of documentation, appropriate and applicable standards, specifications, requirements and or a combination of some or all items	Interviews, observations, reviews of documentation, appropriate and applicable standards, specifications, requirements and or a combination of some or all items applicable	Interviews, observations, reviews of documentation, appropriate and applicable standards, specifications, requirements and or a combination of some or all items applicable

(Continued)

TABLE 4.1 (*Continued*)
Comparison of Audit Processes

Item	Internal Audit	Second-Party Audit	Third-Party Audit
Findings	Reach agreement	Reach agreement	Communicate findings in writing
Closing meeting	Very informal	Informal to very formal depending on the relationship	Very formal and mandatory
Audit report	Informal, based on procedures set by the organization. Often reported on an official form or corrective action report	Informal to very formal; based on agreement with the client. Often reported on a form or a corrective action report	Very formalized and standardized report
Person responsible for identifying corrective actions and closure dates	Manager of the specific area	Auditee's responsibility (the auditor may have input)	

Regardless of the kind of audit, it is imperative that auditors and the lead auditor have no direct or indirect responsibility in the audited area or with the involved personnel. This is so that there is no suspected bias in the findings. The most common types of audits in reference to ISO are described as follows:

- *Conformity assessment*: They are developed for each new approach directive as they are being developed. In general, a conformity assessment examines all activities that assure the conformity of products to a set of standards, including testing, inspection, certification, and quality system assessment. Depending on the products and health and safety they present, the conformity assessment can range from a full-quality assurance system guided by special requirements to manufacturer certification. Because there is a choice, the European Union (EU) considers conformity assessment a voluntary rather than mandatory process, *i.e.* the manufacturer chooses the option.
- *Eco-Management and Audit Scheme (EMAS)*: This audit system was developed to prevent, reduce, and as far as possible eliminate pollution, particularly at its sources. It uses a polluter-pays principle that is similar to the Environmental Protection Agency (EPA) program that permits companies to trade emissions allowances. The binding part of the regulations is that nations must establish methods for companies to validate the conformity and effectiveness of their environmental management systems (EMS). Participation by companies is voluntary. The eco-audit program sets forth

requirements that EU member states participate in an international system of conformity assessment for environmental management. This system is similar to the worldwide traceability system now falling into place for ISO 9001 and is known as ISO 14001.

- *Adequacy audit*: This audit is usually an internal audit known as a *system* or *management audit*. Its function is to determine the extent to which the documentation meets the applicable standard(s) and applicable requirements (industrial and or customer specific).

- *Compliance audit*: This audit is performed by a company that seeks to establish the extent to which its documentation system is implemented and followed by the workforce. The focus is the system or process, not the product. Generally, a compliance audit is related to government regulations or occupational and safety hazards. Not being in compliance is much more serious than not being in conformance. The penalty for not being in compliance may be a financial one or the top management team may end in jail. On the other hand, not being in conformance may result in a verbal warning from your customer, and in the worst case, the organization may lose its business.

- *External audit*: This audit may be an adequacy, conformance, or a compliance audit and is usually performed by a company on its own suppliers or through a third party.

- *Distance audit*: A review of all appropriate and applicable documentation to see if all mandated requirements are met. This review may be conducted in an office, or conference room in the organization's facility (known as the desk audit), or it may be conducted in the third party's facilities (known as distance audit).

- *Extrinsic audit*: This audit may be an adequacy, conformance, or a compliance audit and is usually performed by an independent third party or a customer coming to look at a supplier or a supplier's supplier.

- *Internal audit*: This is perhaps the most common and most important of all audits. ISO 9001 clause 9.2–9.2.2 and IATF 16949 clause 9.2.2.1–9.2.2.4 require a company to audit its own quality system, procedures, and activities in order to establish whether they are adequate and being followed by the workforce. Furthermore, the ISO and the IATF standards call for communicating the results to management as well as for planning corrective and preventive action, if necessary. This type of audit provides excellent communication in the organization and provides management with appropriate and timely information about the quality system and its effectiveness.

- *Layered process audits (LPAs)*: These audits are a quality technique that focuses on observing and validating how products are made, rather than inspecting finished products. LPAs are not confined to the quality department but involve all employees in the auditing process. Shop floor people (operators and/or supervisors) audit process details, daily or weekly. Next level of management audits the audits, weekly or monthly. Top management audits the middle management, monthly or quarterly. The basis for this audit is IATF 16949 section 9.2.2.3.

- *Product/process audit* (based on IATF 16949 sections 9.2.2.3 and 9.2.2.4): This specialized audit examines all the systems that go into the production of a specific end product or service. It is usually called a *vertical audit* and should not be confused with an inspection program of an item. Product audits are applicable to auditing specific projects or contracts, especially in the software and electrical industries. A trained professional auditor should undertake the product/process audit. The auditor may be on company staff, a hired professional, or a third party. What is of paramount importance is that the auditor must be independent and not have direct or indirect responsibility for the audit area or its personnel. Whether the auditor has knowledge in the specific area or activity is not relevant because the auditor's responsibility is to look for objective evidence of conformity based on the requirements of the standard and provided by the documented system.

If the audits are done appropriately, they provide the following benefits to the organization:

- *Management orientation*: An audit will provide information on the current quality system to the management team so that appropriate evaluation can take place. The impetus for this evaluation may be the management team itself, the customer, or the competition.
- *Internal assessment*: An audit conducted as an internal assessment will provide a very good measurement of both efficiency and the effectiveness of the system as well as a strong benchmark for continuous improvement in the future. (Efficiency deals with internal allocation of resources, and effectiveness deals with the level of customer satisfaction.)
- *External assessment*: An audit conducted as an external assessment will establish the credentials of procurement quality standard(s), *i.e.* suitability, conformity, and effectiveness, and in addition will provide supplier certification.

ASSESSMENT METHODS

Each certification body (CB) has its own method of assessment and certification. However, all of them base their assessments on the current documentation of the organization and the site-audit results. The routine events of the certification procedure are as follows:

- *Initial contact from the organization to the Registrar*: General information is exchanged and appointments are set for the preassessment meeting.
- *Preassessment visit*: Either a questionnaire or an actual visit takes place to establish the amount of work needed. For a self-assessment, review Chapter 2.
- *Quotation from the Registrar*: A formal quote for the certification and surveillance services is given to the organization.
- *Acceptance of the quote*: The organization signs the quote or a legal contract for the certification and surveillance procedure as well as the price.

- *Review of the quality system*: The Registrar requests the quality system – sometimes called the documented system – for a review. This review compares the written system (quality manual, procedures, instructions) of the company with the specific ISO standard elements (or industrial and or customer-specific requirements) that the company is seeking certification in. Sometimes, this audit is called a desk audit. The purpose of this audit is to identify any omissions, ambiguities, or major non-conformities prior to the physical (site) compliance audit.
- *Resolution on the desk audit*: When the quality system is accepted, a request for the site compliance audit is made.
- *The compliance audit occurs*: It is generally undertaken with a team of auditors (assessors). The team is made up of one lead auditor and three or four auditors. The time is usually no more than 5 days. A typical audit has the following elements:
  - Opening meeting
  - Audit
  - Closing meeting
  - Reporting results.
- *Certification is issued*: If everything goes well and no NCs or non-compliances are found, then certification is issued. Although the actual certificate arrives 4–8 weeks after the assessment, the organization is now certified.
- *Certification is denied*: Denial is caused by either minor NCs or non-compliances or major NCs or non-compliances. (For certification denial with minor deficiencies in the system, there must be multiple ones (usually more than five) there are exceptions of course). In either case, the organization might be reassessed when ALL the deficiencies are corrected.

Any non-conformity must be recorded and its effect assessed by the lead auditor. Some Registrars have a guidance procedure to assist in the ultimate decision. Let us remember that a major non-compliance or a NC is the absence or the complete breakdown of a required element of the system. A required element is any of the subsections of section 4 of the applicable standard. Sometimes the term major non-compliance is used interchangeably with serious non-compliances or hold points.

On the other hand, a minor non-compliance or NC is an isolated failure to comply with specified requirements. A single minor non-compliance would not normally be reason to deny certification. However, a series of related minor non-conformities may, in the judgment of the lead auditor, constitute a breakdown of a procedure or of the entire system.

At this point, all related minor deficiencies are classified as major. For example, suppose the auditor finds a document without the appropriate signature as defined in the quality manual or a procedure certainly an NC item. The auditor makes a note of it and completes the audit. The auditor finds no more infractions of any kind, reports the lapse as a minor non-compliance (or maybe as an observation), and asks for a follow-up to assure that it does not happen again. On the other hand, if the auditor finds unsigned documentation in other departments of the auditee, that infraction may be viewed as a major non-compliance, since it indicates a system breakdown.

- *Surveillance*: The ongoing program of making sure that the organization maintains the quality system.
- *Appeals*: If during the audit an auditor does not conduct himself/herself in a professional manner or some other complaint is justified by the auditee, the auditee has the right of appeal to the Registrar. The actual process depends on the specific Registrar. The Registrar will investigate the complaint always with an independent third party. If the Registrar is unable to satisfy the customer, the auditee has the right to complain to the CB. The CB may review the complaint by asking for additional information and authorize a new audit. In any case, the decision of the CB is final. The Registrars have the right to decertify an organization that fails to maintain an adequate standard or misuses the logo of the Registrar and the certification.

WHO IS INVOLVED IN AN AUDIT?

A typical ISO or IATF audit requires the following participants:

- *Auditor*: A person who is qualified to conduct the audit in accordance with the ISO and IATF standards.
- *Lead auditor*: Usually the lead auditor is the person who plans the audit and does the preliminary as well as the closing paper work. He or she is a certified auditor and may be called the team leader.
- *Client*: The organization that requests the auditing organization to conduct the audit.
- *Auditee*: The organization to be audited. The entity on the receiving end of the audit activity.
- *Registrar*: The body that issues the certification to the auditee.

The requirements and qualifications for the auditor and lead auditor are lengthy and numerous and beyond the scope of this book. However, the reader can find them in the ASQ (American Society for Quality) document Certification Program for Auditors of Quality Systems (1993). Table 4.2 summarizes some of the auditor's requirements and qualifications.

In addition to specific experiential background and educational requirements, the following aptitudes and attributes are necessary. Good reminders for understanding auditors and their function are the following:

- A third-party organization or individual hired by the client to conduct the audit on its behalf.
- A third-party organization or individuals hired by an approval agency to conduct the audit, initiated at the request of the client.
- Second-party auditors, employed by the customer, potential customer, or other independent organization requesting the audit of the auditee.
- Second-party auditors, employed by an approval agency, carrying out an audit to determine the ability of the auditee to provide the desired quality system, product, service, or process.

TABLE 4.2
Summary of Key Requirements for an Auditor

General Aptitude	Technical Aptitude	Personal Attributes
Qualifications	Knowledge of technical standards	Leadership
Certification	Knowledge of cost accounting	Interfacing abilities, confidence,
CQE; CQA		composure, independence
Knowledge of	Knowledge of quality cost	Planning abilities, understanding
standards and or	systems	(industry, process, product, quality
requirements		system, investigative techniques)
Knowledge of	Knowledge of statistical	Communication abilities (oral and
management	techniques (sampling,	written)
systems and styles	frequency, analysis, inferences)	
Ethics	Knowledge diagnostic	Critiquing ability
	techniques., problem solving	
Integrity	Technical troubleshooting	Decisiveness

CQE = Certified Quality Engineer; CQA = Certified Quality Auditor.
Both are specific certifications offered by American Society of Quality (ASQ).
IRCA = International Register of Certificated Auditors.

- Auditors, employed by a corporate headquarters, carrying out an audit to determine whether a division or other element of that corporation complies with corporate policies and desires.

On the other hand, the term internal quality auditor refers to auditors who are employees of the organization being audited.

AUDITOR

An auditor is the person who evaluates the QMS for an organization for its effectiveness. The evaluation is against standards and/or requirements. An auditor may be certified or not certified. The current trend for all auditors (internal and external) is that they have to be certified. Certification is granted by ASQ, which establishes certification standards, administers certification examinations, and offers preparatory classes for the exams. Requirements for certification include 8 years of work experience in quality control, three of which must be in a position of authority over quality processes. A high school diploma is the minimum education prerequisite, though further education can be substituted for years of work experience. Certification also requires passing a 5-hour, open-note exam that is offered twice a year by ASQ.

Sometimes, quality auditors conduct tests on a company's products to ensure that the products look and act like they are supposed to. Professional certification by ASO requires a certain amount of work experience, along with passing an exam.

LEAD AUDITORS

The basic requirements for a lead auditor are the same as for the auditor, but there are additional experience and training requirements for the lead auditor. The lead auditor must have at least 25 additional total audit days than the auditor and complete a 36-hour training program sanctioned by a certifying body. This requirement depends on the CB, the context of the audit, and may vary substantially from CB to CB.

CLIENT AND AUDITEE

No specific requirements except the desire to participate.

REGISTRAR

Extensive requirements by the ISO Central Committee and the individual country of participation. In the United States contact the Registrar Accreditation Board (RAB) which maintains the latest requirements.

TYPES OF AUDITS

Types of audits vary. However, all audits are classified as either first, second, or third party. The first-party audits (internal audits) are done internally or by a sister plant. First-party audits can also be outsourced to an outside auditor. Customers for various reasons conduct second-party audits with their suppliers. Outside registration firm or government agency conduct third-party audits to award various certifications.

INTERNAL AUDITS (SELF-AUDITS)

As a requirement of most customer-specific requirements and internal quality standards, organizations are conducting internal audits. The purpose of these audits is to find weaknesses in the management systems. Audits are to be planned and carried out in a systematic manner. Internal audits rely on competent auditors, which normally is demonstrated with training and evaluation of the auditor. Internal auditors are often recruited from various areas within the organization. It is important that these auditors must not audit their own environments. The reason for this exclusion is to eliminate bias in their finding. After all, the purpose of this audit is not to find old forms or a missing record(s); although those are valid issues. The real purpose should be to drive value back into the organization. Auditors have a tremendous opportunity when conducting internal audits. They have the opportunity to help the organization become stronger by identifying best practices, areas of waste, and non-value-added activities. The level of skills and background of the internal auditor as well as management's commitment influence the approach.

Organizations have a choice with how often they audit their processes and who can carry out the audit. The only uncompromising rule for an internal audit is that the auditor never should audit his or her own work. Naturally this is a conflict of interest. The frequency of internal audits is normally planned on a master schedule or calendar. Typical organizations may plan for one per year. This is not the most value-added approach but certainly one can argue will satisfy a customer or third-party Registrar.

The effectiveness of these audits depends on several factors: auditor skills, planning, and management commitment. Skills are developed overtime, which implies that in order to be a good auditor, it will take some experience in conducting audits. Experience will improve your skill of asking questions, but more importantly it will eliminate the fear of asking probing questions. Good audit planning will help in becoming competent in auditing. On the other hand, if the planning is done haphazardly or without consideration to the auditee and process, it will impact the effectiveness of the audit program in a negative way.

Management commitment to the internal audit program is vitally important (ISO 9001 clause 5.1 and IATF 16949 clause 5.1.1.1–5.1.1.2). The level of support that is provided and resources allocated are all elements of management commitment. Audit programs need competent and trained people to carry out audits. The management teams that believe in the value of audits will provide the appropriate support.

Second-Party Audits (Customer Audits the Supplier)

Customers often like to assess their suppliers. There are many reasons that customers audit suppliers, such as, conformity to supplier management approval processes, sourcing decisions or adherence, or ability to meet their requirements. The structure of these audits varies. They range from checklist-driven audits to process audits. Many organizations will have some defined format that they will share with the supplier. Others may not. The best advice is to take time to review and agree upon the expected methodology (tools and criteria) to be used so that to potentially circumvent any confusion or misunderstanding during the audit. It is imperative that the customer does not think that they have super power of the supplier and bully them around. If that happens the relationship will not produce fruitful results. If anything, it will breed resentment and the audit will not produce value-added benefits.

Third-Party Assessment (A Third Party Audits the Organization's QMS)

A third-party assessment is undertaken by an independent body to establish the extent to which an organization meets the requirements of an applicable standard and/or set of regulations and or customer specifications. Third-party assessment bodies can assess against any required standard. The independent auditing body – the *Registrar* – issues a certificate of registration that indicates acceptance of the organization as a company of assessed capability or something similar. The Registrar issues this certification after a review of the quality system and a physical audit of the organization. The original certification takes approximately 10–24 months (from preparation to submitting the application to certification). After the certification is achieved, the CB visits the assessed organization once or twice a year for surveillance purposes and every 3 years for a major recertification assessment. Generally, the cycle of surveillance and recertification depends upon the policy of the Registrar. On the other hand, generally, customer specifics require a surveillance once a year (and if there are problems, they may be more frequent). An example of this is the manufacturing site assessment (MSA) for the automotive industry.

The certification bears witness that the assessed organization complies with all of the requirements of the applicable standard. However, certification does not guarantee product or service quality. Perhaps this is the area where much more communication is needed to make sure that everyone recognizes that audits are system oriented and not product or service. It guarantees only that a quality system has been defined and exist.

AUDIT PLANNING

Audit plans serve several functions during an audit, some of which are listed below:

- First, the audit plan helps to structure the audit by identifying the areas to be evaluated, the sequence of evaluation of these areas, and the approximate time that will be spent in each area.
- Second, the audit plan can be used to clearly define responsibilities among an audit team by associating each audit area to an auditor who will lead the questions in that area.
- Third, the audit plan communicates to those who will be interviewed when they will need to be available. This allows them to plan the rest of their day and helps ensure the availability of these key individuals.
- Finally, the audit plan is also normally used to indicate audit events such as when and where the opening and closing meetings will be held and when the auditors will break for lunch, as well as other logistical issues as needed (*i.e.* breaks, conferences).

The audit plan, if used, should be developed and communicated to the auditee management at least a week prior to the start of the audit.

In real estate, it is said that, location, location, location is a key factor. In auditing, the key characteristic of success is: planning, planning, and planning. As is true in most activities, thorough planning is needed to ensure that the audit will be successful in finding the weaknesses and opportunities for improvement that exist within the management system. Planning is vital for all auditors; new, part-time, or for the person auditing outside of their area or process.

Unfortunately, planning, more than any other audit activity, is often done haphazardly if at all. Remember, the primary purpose of the internal auditor is to find weaknesses and opportunities for improvement. On the other hand, the purpose of the second- and third-party audits is to verify that a system exists and it is followed. Therefore, a successful audit is a function of the auditor having invested time in planning the audit. This is essential because:

- Planning ensures that everyone agrees with the purpose and scope of the audit.

- Planning exposes the auditor to the processes and procedures that will form the basis of the audit.
- Planning allows the auditor to develop a strategy for the audit.

In addition to these, appropriate planning will crystalize the scope of the audit so that everyone agrees as to what the purpose is and what should be considered. Typical items for consideration under the scope discussion should be

- Organizational units or departments to be audited
- Physical locations or areas to be included in the audit
- Specific activities and processes to be evaluated
- The timeframe (length) of the audit
- Other considerations may be
  - Document and record control
  - Training
  - Maintenance
  - Continual improvement, including quality objectives
  - Control of monitoring and measurement equipment
  - Quality planning.

The most effective quality system audits are typically those that evaluate an entire value chain, from receipt of order to delivery for a specific product. The challenge with optimizing the scope is *time*. To audit an entire value stream might take an auditor 2 or even 3 days (even more days, if the organization is large or has multiple facilities). Add on an additional 1 or 2 days' worth of planning time and you can quickly see the magnitude of this problem, which is especially acute when the organization uses part-time auditors. To address these challenges, audit program managers often lessen the scope by splitting these larger audits into a smaller number of focused audits. Audit program managers must be careful in this balancing of needs versus resources, since splitting the audit into ever smaller chunks impacts on the ability to test the linkages and alignment between interdependent processes, which is where experience has shown most systematic weaknesses exist.

So, how do we make sure that the audit will be effective in evaluating the quality system of the organization? The answer is a resounding one word, *planning*. This planning is materialized by generating a *checklist* that will guide the auditor through the process in a systematic approach.

Of course, to do the planning and have it articulated with a checklist, one must understand the process, that is, the inputs, the transformation (value-added task), and the output. All that information is part of (a) quality manual – if it exists, (b) procedures, and (c) instructions. Given this documentation, the auditor asks pertinent questions to validate the existence of the management operating system (MOS). In no way however, the checklist becomes an absolute guide of questions. It is not unusual as part of the audit and based on the response of the auditee to develop a *new trail* of questions which may not be part of the original questionnaire. If the checklist becomes an absolute crutch for the audit without flexibility, it is guaranteed that fruitful information will not be generated. If that is the case, audits become *sterile instruments of passive inspection* and the weaknesses of the system will be rarely identified, if ever.

It is important here to recognize that an auditor DOES NOT have to be an expert on the process to be a good auditor. Rather, it is important to understand it well enough to ask the right questions and evaluate the adequacy of the responses. So, the key to understanding the process is to become familiar with all the pertinent documentation of the process for review, *i.e.* procedures, instructions, flowcharts, and so on. If a flowchart is available, make sure as an auditor that the inputs and outputs are clearly identified. If not available, then ask (strongly suggest) the auditee to generate one. The information gained from this review will be the basis for the audit strategy. This review is called *desk audit* and may be conducted through a *distance audit*. Based on the review, an auditor may zero-in on specific items and deep dive into them for more specific information. Of course, the detailed information will demand more time and flexibility, and as such, the auditor must be cognizant of the time allowed for completion. It is imperative, however, to audit good, poor, and marginal processes, if they are known to be so. We are interested in an unbiased output, therefore through appropriate sampling, such as *stratified sampling*, may be used to avoid false responses. (Under normal situations, *rational sampling* works fine.) In a sense, here we suggest that the auditor becomes a *trouble shooter*. This is because the NCs found through questioning, more often than not, are caused by management system failures and they have to be identified and communicated. Auditing is one way to accomplish this.

The best audit checklists not only include the questions to ask and record on a document the findings, they also provide areas for comments and information about the samples selected, documents reviewed, *etc.* In fact, a well-constructed checklist will also guide (by its flow of questions) the auditor in how to verify the response by providing defined areas for the *objective evidence* to be examined. An effective checklist is the one with open questions as opposed to closed ones with an expected response of a Yes or No. An open question will give you more information and create a conversation with explanations that otherwise will not be available. A generic guide of asking questions is the following path: Is the question jargon free? Does one question lead into the next? Can the objective evidence reviewed for this question help the auditor select a sample to use in verifying the answer for follow-up questions?

One of the greatest benefits of a checklist is that it can help structure the audit. The essence of an audit plan is to provide a "macro-structure" by identifying the broad areas to be examined and allocating time for each area. A good audit checklist should provide a structure for each individual area or process to be examined. It does this by including questions relating to requirements that need to be verified in the approximate order in which they will be asked. We use the word *approximate*, because the auditor may need (in fact, quite often) to jump ahead or back on the checklist to adjust the audit flow to the auditee or the natural process or because the question and/ or observation may create a new *audit trail* in investigating a concern. In general, the questions on the checklist should follow the natural flow of the process. Since most procedures are written this way, audit checklists generally follow the process flow. The presentation of audit evidence may sometimes warrant departing from the natural flow to pursue an area relating to a record, list, or other object.

A word of caution: Checklists only capture requirements from the applicable standards or specifications. They do not capture all of the additional requirements that your organization calls out in its detailed procedures and instructions, nor do they adequately display all of the important inputs, controls, and resources into these processes that are unique to a company. They serve as a solid foundation to prepare from, not as a ready-made, complete checklist that an auditor can just pick up and use. The audit program manager and lead auditors should modify these checklists to include these additional internal requirements and inputs, using your procedures, instructions, Value Stream Mapping (VSM), Supplier-Input-Process-Outputs-Customers (SIPOC) diagrams, flowcharts and so on. These internal requirements should replace some of the generic requirements on the checklist. (Special note: While the pre-defined checklists for IATF, VDA, and some of the ISO standards are good, they may not cover all the specifics for an organization. Therefore, the lead auditor should review them very closely and add – as appropriate – additional questions to reflect the specific needs.)

If the audit program manager chooses to use pre-prepared checklists, then the auditor must take extra precautions to ensure that the specific requirements for the organization are accounted for. These precautions could include providing copies of relevant procedures and instructions to the auditors (rather than making them find and copy them themselves), having the auditor present his strategy or reviewing the audit plan prior to the audit.

AUDIT CRITERIA

Just like everything else, success depends on what we mean with certain words. In auditing, we have the same problem. So, let us examine the key words that makes an audit significant for all concerned. The words are: *Audit criteria*, which mean – standards, policies, practices, procedures, or requirements against which the auditor compares collected *audit evidence* about the subject matter. Keep in mind that a good amount of the third-party auditor's time will be spent evaluating compliance to customer-specific requirements. Internal auditors should do likewise.

In other words, **Audit Criteria = Requirements to Verify:**

- Specification (ISO 9001, IATF 16949, *etc.*)
- Documented procedures (if they exist)
- Industry requirements, if any
- Customer-specific requirements (for audits of product or service realization)
  - Specifications, contract requirements, drawings
  - Customer-specific requirements on International Automotive Oversight Board (www.iaob.org)

Serious consideration should also be given to the following:

- Previous audit results for this process
- Significant customer complaints or problems
- Quality objectives established for this process

- Performance metrics (they will help developing the "micro-scope")
- Review actions if performance is not achieving planned results.

AUDIT FINDINGS

To deliver value through the audit process, the audit team members have to ask "the" tough questions and conduct *a deep-dive* into the process in order to deliver value. Auditors can make a difference with findings and opportunities that are noted during audits. Auditors should never be intimidated and discouraged by the process owner(s) from writing findings or opportunities. The findings fall into the following categories:

- Verification of conformance
- Evaluation of effectiveness
- Identification of opportunities for improvement
- Identification of best practices.

A fundamental purpose of any audit is to verify that the organization is conforming to the relevant standard and to its own internal policies and procedures. A management system cannot produce results if it is not followed. Non-compliance or NC is a weakness that will lead to inferior performance of the management system and such instances must be identified during the audit. On the other hand, conformance or compliance means that the organization is adhering to the requirements set forth in its internal procedures, policies, guidelines, and to external requirements set forth by the specification, its customers, and/or adopted industry practices. Basically, it verifies that we are "doing what we say we do." The following are examples of NC issues:

- Failure to comply with the standard(s)
- Failure to comply with specific customer requirements
- Failure to comply with regulatory requirements
- Failure to comply with internal policies and procedures.

Conformance issues are normally cited as non-conformance, or NCs. Sometimes they are classified as major and minor. A major NC is the absence of a required procedure or the total breakdown of a procedure or process – system failure. A minor NC is a single observed lapse in a procedure or process. A number of minor nonconformities listed against the same requirement can represent a total breakdown of a system and thus be considered a major non-conformity ("The Trivial Many").

A secondary purpose of the audit findings is to verify the effectiveness of the process. This is accomplished by questions, such as: (a) is the process achieving planned results? (b) If not, are actions being taken to achieve them? Therefore, if the system is not effective, it may result in unfavorable NCs.

AUDIT SCHEDULE

Even though the outputs from these processes, and therefore information relating to their performance, will be continually reviewed during other audits, there are still centralized functions that may not be examined without a focused review.

These audits should focus on these centralized functions only. Time should not be wasted pulling documents from the floor to review revision level control or checking the calibration of production measuring equipment, since these *should have been reviewed* adequately during other process-based audits.

An audit schedule does more than simply identify the dates when quality audits will be conducted. An audit schedule also reflects the audit program manager's strategy for performing management system audits. Indeed, the development of the annual audit schedule is in reality a strategic action. Therefore, the audit schedule is important because it sets the overall scope of the audits. By scope we mean what processes and/or areas will be included in each audit and how much time is allotted.

Let us all remember that the most effective audits are those that test the linkages (interfaces) and handoffs between related processes. Combining multiple interdependent processes in the audit scope allows the auditor the opportunity to test these linkages. The most effective quality systems audits are typically those that evaluate an entire value chain, from receipt of order to delivery for a specific product.

PROCESS FOR DEVELOPING COMPLIANCE

The minimum requirements to start any audit fall into three categories. They are as follows:

STARTING

Knowing what to do: An auditor must gain some knowledge about the organization and the audit environment before starting the physical audit. The following recommendations will help in the preparation phase:

- Gathering and collating information, policies, procedures, and standard operating procedures. Review the appropriate and applicable checklist (if prepared).
- Developing additional documents to facilitate the audit. Checklists and work notes may prove worthwhile. Work teams and other resources may be applicable.
- Defining and recognizing the responsibilities and authorities, particularly those stated in ISO standards, the industry (IATF), and the specific organization requirements.
- Communicating with all workers and certainly the worker(s) responsible for the audit. Audits should not be planned as surprises (with an attitude of "I'll get you"). Assumptions should not be made. Communication lines can be kept open with telephone calls, questionnaires, reconnaissance visits, e-mails, conference calls, or any other means that are applicable and appropriate for the situation.

DOCUMENTING

Writing it down: An auditor has the responsibility to be prepared, fair, impartial, inquisitive, honest, and observant. To do all that, some of the following recommendations may help:

- Writing down everything seen at the time of the audit and having the supervisor or operator sign the notes. This practice will eliminate confrontational statements later on during the reporting of the audit. Make sure you are able to separate "observations" from "conclusions."
- Developing (or review for accuracy) process flow diagrams with parallel information dealing with responsibilities, authorities, and special instructions. This information will help an auditor to prepare for the audit and to understand the process. Although a process flow diagram is not essential to conduct the audit, it does help the auditor conduct a cursory comparative analysis of the process at hand and the organization. Even more beneficial will be the presence of a Value Process Mapping (VPM) – if available. When used in conjunction with the checklist, the process flow diagram and VPM becomes a very powerful tool indeed.
- Communicate with all workers and certainly the workers affected by the outcome of the audit. Do not plan for any surprises. (However, surprises may surface for both the auditee and the auditor.) Be open, concise, and easy to understand. Avoid jargon and traps of any kind.

EVALUATING

Self versus independent: An auditor's ultimate responsibility is to evaluate the quality system of a given organization, based on the organization's appropriate standards and follow-up – if this is a surveillance audit. The following suggestions may help in the evaluation process:

- *Knowledge of applicable standards*: If the auditor is not familiar with the organization and the audit standards, the evaluation may suffer.
- *Independence of unit(s) assessed*: As we have already discussed, the independence of the auditor is of paramount importance. Without this independence, bias and doubt enter into the evaluation.
- *Experience in assessment(s)*: We addressed the issue of area-specific knowledge versus auditing knowledge earlier in this chapter. However, we must emphasize the importance of auditing knowledge. Unless the auditor is experienced in internal and external audits, the evaluation may omit essential issues and concerns. The following factors require experience and audit knowledge:
  - Recognition of a contractual requirement to audit.
  - Investigative procedures for actual or potential problems.
  - The appropriate audit for the appropriate standard.

- Auditing a shutdown department or the records of a discontinued supplier.
- Recognition of the life cycle of the contract or project.

Consideration and evaluation of these factors will indicate to the auditor whether or not an audit of the specific activity is required or whether that activity is worth pursuing. Depending on the evaluation, the auditor may select the appropriate type of audit and may define or redefine the schedule of the audit.

Based on this evaluation, the auditor may define the type of audit (system, compliance, product, *etc.*), redefine the schedule of the audit, or change the sequence of the audit. What is more important is that the auditor will plan for the best – and most appropriate – audit and identify the scope of the audit, dates and schedule of the audit, and who will perform the audit.

In summary, an audit is a documented activity performed in accordance with written procedures or checklists to verify, by evaluation of objective evidence, that applicable elements of a quality assurance program have been developed, documented, and effectively implemented in accordance with specified requirements. It must always be remembered that the documented and written definitions of activities are in the domain of the management and that the auditor is looking for objective (replicated) evidence that quality assurance (quality planning) has been developed and implemented.

AUDIT PROCESS

All audits follow a sequential path that has four basic phases. They are as follows:

- Phase 1 *Preparation (preaudit)*: This phase includes selecting the team, planning the audit, and gathering the information.
- Phase 2 *Performance (on-site visit)*: This phase begins with the opening meeting and finishes with the actual audit.
- Phase 3 *Reporting (post audit)*: This phase includes the exit meeting and the audit report.
- Phase 4 *Closure (review the results (good and bad) of the audit)*: This phase includes the actions resulting from the report and the documentation package.

Phase 1 – Preparation

The planning activities depend on the organization and the certification it seeks. They are usually quite extensive and beyond the scope of this book. Some good sources for further information on the topic of planning for audits are Arter (1994), Hutchins (1992), Robinson (1992) Mills (1989a, 1989b), Sayle (1988), Duhan (1979), and Emmons (1977). Here we are identifying only the most typical planning activities.

The first planning activity is determining the implications of the quality audit:

1. Identifying the purpose of the audit. What do the customers want? Is the control system adequate? Is the control system working? Is the control system used by management?

2. Identifying the scope of the audit. What are the boundaries? Are they determined by product line? By process area? By customer? By systems? By organization? For ISO certification, the boundaries are generally set by the facility.

3. Identifying the resources and defining the audit team requirements. How many members will the team have? One person will invite bias and more than seven will create a mob. From the experience of this author, the optimum number of team members is four or five. Another consideration in the selection of the team is the composition. Care should be taken to have trained and qualified auditors who have both auditing experience and knowledge of the process to be audited.

4. Identifying the authority and defining the authority so that the auditor can establish legitimacy in the course of the audit, defuse hostility, avoid argumentation, and minimize – if not completely avoid – wasting time.

5. Identifying the initial contacts. When meeting with the auditee make sure some of the following items are under consideration. Set the date (be firm but remain flexible). Review both the purpose and scope. Make sure you know the names and the titles of the people you are to meet with. In case of formal contact by letter, that contact ought to be through the auditee's purchasing department or the executive department. The introductory letter should contain the following information:
   a. Auditee name
   b. Purpose scope
   c. Appropriate standard activities to be audited
   d. Applicable background documents
   e. Identification of team members
   f. Preliminary schedule.

6. Determining the use of checklists. The auditor is responsible for ensuring that the audit flows smoothly and makes sense. A checklist
   a. Provides structure in the audit.
   b. Assures that the audit will cover the required areas.
   c. Provides communication.
   d. Provides a place to record findings (good and bad).
   e. Helps in time management.
      (The checklist should not stifle the creativity of the auditor, limit the flexibility of audit, or confuse the applicability of the audit. Rather, it is a guide that facilitates and NOT a prescription for specificity. It may aid in identifying *trails* which may take one completely off the official checklist.)

7. Checking different sources for clues. The more information the auditor has, the better the chances of successfully evaluating the quality system. The scope of evidence includes the following:
   • Physical evidence (current and historical).
     – Reinspect, retest, recheck, and rejects (waste).
     – Laboratory results, calibrations.
     – Any recorded documentation.

- Sensory observations (understand activity).
  - − Observe tasks being performed (online).
  - − Visual checks of products.
  - − Storage of parts.
- Documents and records.
- Interviews and questions of personnel performing, storing, analyzing, and evaluating data. Because of the nature of interviewing, the responses should be corroborated by either another person, document, or direct observation.

The following activities are also part of the preparation phase.

- Understanding the resources needed for the audit. In every audit, there are several items that require appropriate planning to ensure success. Some critical elements are personnel, office facilities, plant facilities, sequencing, scheduling, and working papers. Some of the background knowledge that will help ensure success is as follows:
  - Memory prompters for the auditor.
  - A comprehensive list to allow for grading of the effectiveness on specific element rather than a check-off sheet.
  - Detailed records of what was audited and where the activities of the audit were carried out, plus the findings.
  - Applicable statistical techniques.
  - Sampling and appropriate evaluation techniques.
- Scheduling the audit. How and when is the scheduling to be done?
- Sequencing the quality audit functions. Is the sequence of the operations important? Will the results be different if the sequence is not followed?
- Preparing or gathering working papers for the audit. These are all the documents required for an effective and orderly execution of the audit plan. By format and content, they describe the scope and approach of the audit assignment and its operational elements. Specifically, they define the tools that are appropriate for use during the audit and applicable to the organization. Some of the tools and documents that the auditor may use or find during the audit are as follows:
  - Process flowchart.
  - Failure mode and effect analysis.
  - Control plan.
  - Gantt chart.
  - Critical path network.
  - Matrix responsibility chart.
  - Cause-and-effect diagrams.
  - Appropriate statistics.
  - Planning documents.
  - Checklists.
  - Reporting documents.
  - Procedures.

- Determining the sampling procedures to be used in the audit. What is appropriate sampling for the processes audited? Is the frequency appropriate and applicable?
- Deciding how the audit observations are going to be interpreted.
- Deciding how to report the results of the audit.
- Preparing procedures for requesting and following up on corrective action.
- Preparing appropriate and applicable prevention approaches – if possible.
- What are the results (product) of the preparation phase? There are at least four specific outcomes at the end of the preparation phase.
  - Creation of the audit plan.
  - Creation of the specific checklist questions.
  - Creation of an action plan for the specifics of the audit.
  - Creation of possible ideas of preliminary evaluation.

Phase 2 – Performance

After all the background information has been completed, the audit team is ready to conduct the audit. There are two stages in the audit evaluation.

1. The manual versus the requirements audit.
2. The physical audit versus the organization's documentation.

In the manual versus the requirements audit (called a desk audit), the auditor thoroughly compares the documentation (generally the quality manual; sometimes the procedures as well) with the standard of choice seeking ISO 9001, 14001, 18001/45001, VDA, or IATF certification. If everything is in proper order, the audit will proceed to the second stage. If the review finds NCs in the documentation, appropriate corrections and prevention steps must be made before the second stage can take place. The desk audit usually is performed away from the site, quite often in the auditor's or Registrar's office. The desk audit may be included in the preparation phase.

In the physical audit versus the documentation (called the site visit or site audit), the auditor thoroughly compares what actually occurs to what the documentation says will occur. Even though the physical audit is quite detailed, it is only a random snapshot, or sample of the system, in relationship to the entire organization. A typical performance phase is made up of four stages. They are as follows:

1. Opening meeting
2. Understanding the controls
3. Verifying/validating the system
4. Sharing information.

Opening Meeting

The opening meeting in any audit sequence is generally viewed as an introduction of all parties involved. It is mandatory, and all appropriate individuals must be present. The meeting should not last more than 1 hour. The typical events of an opening meeting are as follows:

- Introduction of the lead auditor and the individuals present.
- Company presentation (a short overview of the company).
- Confirmation of the standard to be used.
- Confirmation of all previous agreements as to scope and objective of the audit. Sometimes verification and validation of past findings in both internal and or external are also discussed.
- Confirmation and distribution of schedule and acceptance of audit.
- Confirmation of guide and arrangements such as phone, office, computer, logistical support, and meals.
- Confirmation of the documentation system that will be used during the audit.
- Confirmation that the organization is ready for the audit and that all staff and employees have been notified.
- Distribution of the checklist (optional).
- Explanation of the timing and purpose of the closing meeting.
- Invitation of any questions regarding the audit.

At the end of the opening meeting, the audit team goes forth with the site audit or adjourns to its designated office for a last-minute briefing and to arrange times to reassemble for reviews, a meeting with the organization's representative, or both.

Understanding the Controls

In most – if not all – organizations, there are two levels of systems: the formal control system, which is documented, and the informal, which is not documented. The auditor must audit both systems and make sure both systems are followed appropriately and completely.

It is important to remember that the ISO 9001 clause 7.5.3 emphatically establishes that all quality records must be documented. However, the ISO 9001 clauses 4.2 and 4.3 suggest that in some cases the documentation may be informal. Therefore, even though we strongly recommend that the quality system (all documentation) be formal, we recognize and the standard recognizes the possibility that informal controls do exist and are legitimate under ISO. Conducting an audit of an informal system is very demanding, time consuming, and costly. An informal audit gives heavy emphasis to corroboration, personal interviews, and observations.

Verifying/Validating the System

There are as many verification systems as there are auditors. However, all of them present themselves in a general methodology.

> *Step 1. Putting the auditee at ease*: Your presence is threatening. You are the outsider. Make sure you as an auditor defuse a possible difficult situation. You are looked upon as a spy and a person not to be trusted. Be prepared to answer questions that directly or indirectly ask, "How dare you question my process? What do you know of me or my process? What are you going to do with this information?"

As an auditor, you are looking for information that is controlled by someone who may or may not be friendly to the audit process. What do you do? Some of the following hints may help:

- Dress for the occasion.
- Talk in an appropriate manner.
- Ask nonthreatening questions.
- Be firm but informal.
- Demonstrate professionalism. Make the person comfortable.

Step 2. Explaining the purpose of the audit: This second step is perhaps the foundation of building good rapport with the operator. Typical actions are as follows:

- Introduce yourself.
- Demonstrate competence.
- Explain why you are here.
- Show respect for the operator.
- Explain what you are going to do with the information. Emphasize the process not the person.
- Show that you are aware of the system, but don't show off your expertise.
- Show your appreciation of the operator.

Step 3. Asking the workers what they do: An auditor obtains information during an audit by reading, observing, and listening. The auditor also needs to ask questions. To gather as much information as possible, the following suggestions may help:

- Ask questions that are open ended and never personal or threatening.
- Avoid leading questions.
- Avoid questions that start with "Why." They may imply criticism or disapproval.
- Avoid questions that start with "Who." They infer guilt and the person who is asked will begin a very defensive posture.
- Avoid questions that start with "I understand that you."
- Avoid tape recorders.
- Focus on your questions. If you lose your train of thought refer to your prepared checklist.
- Avoid questions with long introductions.
- Avoid apologizing for your questions. It demonstrates insecurity of what you know.
- Ask for forms, records, procedures.
- Wait for the response when you ask the question.
- Control your idle talking; listen twice as much as you talk.

It is quite possible that workers may withhold, distort, or give incomplete information, either by accident or deliberate action. Being aware of this, the auditor needs to use excellent communication skills to ensure that all information received is valid and can be substantiated.

One of the many things that an auditor cannot dismiss when asking questions is the body language of the auditee. Body language is the name given to all those unconscious gestures, facial expressions, or body movements that are part of every personality. The study of body language is the new science of *kinetics*. Warning!!! When trying to interpret body language, remember that different nationalities, racial groups, *etc.* have different codes in this unwritten language. A good auditor is aware of this diversity and will limit body language interpretation to the environment he or she knows.

Step 4. Analyzing what the workers said: In this step, the auditor analyzes all the information from the audit and writes the information in a notebook. The auditor must make sure to write only objective observations and data from the audit, not interpretations of suspicions or innuendoes. To maintain the integrity of the analysis, it is strongly recommended that the auditor ask either the operator or the department supervisor to initiate the notes.

To make sure that the analysis is fair and appropriate, the auditor must address two fundamental questions:

1. Is the control system implemented?
2. Is the control system working?

To answer these questions, the auditor (quite often) must look beyond the responses of the eye-to-eye questions of step 3. There are at least three primary methods used for verification or validation: tracing, sampling, and corroboration.

1. *Tracing*: Is an approach to auditing that traces something through all of the process steps. In the forward mode, tracing starts at the beginning (order receipt) and works forward until the needed information is found. In the backward mode, tracing begins at the end (order completion) and works backward until the needed information is found. The middle mode works both backward and forward from *some* critical point in the process. In most cases, tracing is constantly utilized throughout the audit by all auditors.
2. *Sampling*: Is trying to find the truth in a limited amount of time. Therefore, the more an auditor knows about sampling techniques, the better the audit will go. Some of the basic questions and concerns in regard to sampling techniques in auditing follow:
   • What to sample? The answer will depend upon the critical items or the area under consideration. Specific items that indicate the importance of sampling are depended on at least the following:
     a. Important to the auditor.
     b. Important to the auditee.
     c. Overloaded areas.
     d. Areas with historical problems.
   • What will the results mean? Is the answer significant for final disposition? Is the information incomplete? Are the results acceptable? Why or why not? Always remember that *important* is not the same as *significant*. Make sure everyone understands the difference.

- How to sample? All sampling for any audit must be done on a *random* basis. Unless the sample is random, the results are meaningless because of bias or some other compromise in data selection.
- How much to sample? Again, some knowledge of basic statistics and sampling may help the auditor define the sample. A general rule is that as the number of discrepancies in a population becomes small, the sample size must get bigger to have the same degree of confidence.

3. *Corroboration ensures data integrity*: We know that people draw different conclusions from the same facts. We also know that every individual rank important issues differently. As we mentioned earlier, some operators willfully or accidentally recollect events differently. The job of the auditor is to maintain the integrity of the audit so that the results will be useful to everyone concerned. Therefore, the auditor must:

- Persuade the auditee that the auditor's perception of the facts is better or more useful than the operator's perception of the facts.
- Demonstrate to the auditee the benefits of sharing the facts accurately.

If the facts are so important and at times so elusive, how can an auditor establish the truth? It is the responsibility of the auditor to check and recheck the facts with whatever means are applicable and appropriate under the situation. Typically, the auditor will resort to corroboration when the facts don't agree or something is just not right. Corroboration may be obtained from

- Two or more different auditors.
- Two or more different records.
- Two or more different interviews.
- Any combination of the above.

An auditor should remember that a statement made during an interview is not a fact until it is corroborated by someone else or verified by a document.

If the intent of the audit is to see how many non-compliances or NCs one can find in the quality system, then identify everything. Remember, however, that nothing is perfect. On the other hand, if the intent of the audit is to help the organization improve and establish a good benchmark of their quality system, then follow the Pareto (80:20) principle and do not nitpick. The idea is to focus on chronic or persistent problems or specific trends.

4. *Sharing Information*: At the end of the physical audit, the auditors meet and review the events of the audit (good and bad). They discuss good practices as well as problems, concerns, and specific questions so that everyone on the audit team has a good idea of what went on in all the areas. Another reason for this meeting is to reach a consensus about the disposition of certain non-compliances or NCs. Quite often, the auditee makes minor adjustments based on the audit information. Sometimes the auditee will be briefed as to the status of the audit, pending final results, and the resolution of special concerns and problems.

What is the result (product) of phase 2? In this phase, the audit team collects data about the quality system and relates them to actual events. Phase 2 generates the objective evidence by which the ultimate decision of the audit is going to be made. Objective evidence may be qualitative or quantitative information, records, or statements of fact (corroborated) pertaining to the quality of an item or service or to the existence and implementations of a quality system element or documented requirement that is based on observation, measurement, or testing and that can be verified.

Phase 3 – Reporting

The saying "I know that you believe YOU understand what you think I said, but I am not sure that what you heard is not what I meant" is precisely why the reporting phase is necessary. An auditor wants to pass on honest and accurate information. After all, the main purpose of the audit is to collect information in factual form and to pass it on as objectively without any bias. In this phase, we are focusing on the passing on of that information. The two issues of concern are the reporting and the closing meeting.

1. *The Reporting*: The reports are short and contain the findings of the audit. A third-party audit usually states the findings as NC to specific clauses of the certifiable standard. On the other hand, a first- or second-party audit, in addition to stating the NCs or non-compliances to the standard, may add recommendations to fix these deviations or to improve the quality system. The deviations are based on knowledge and objective evidence from the audit and hardly ever number more than ten. The reason for this low number is twofold. The first is the auditee has gone through at least one preassessment and very detailed preparation. The probability of having more than ten gaps (deviations) from the requirements is indeed very small. The second one is that the auditor who issues more than ten non-compliances may be nitpicking. Furthermore, more than ten deficiencies can create management overload.

   Unless the audit is internal, the auditor's findings and recommendations should not require specific problem-solving actions. In addition, the auditor's findings and recommendations should not be based on perceived bias or gut feelings. Such findings are likely to lack legitimacy and credibility. When ready to write the report, the auditor must keep in mind some very common definitions, such as:

   The Audit Corrective Action Report (ACAR) is the report that the auditor uses to document the findings during the audit. Two generic ACAR forms are shown in Tables 4.3 and 4.4. The report form in Table 4.3 is used on an internal audit; the report form in Table 4.4 is used in an external audit. The difference between the two is that the internal report form has room for recommendations. Sometimes these reports are called NC or non-compliance reports. If an ACAR is issued, the auditor has to evaluate the effectiveness of the corrective action before the audit is closed.

   A major, non-compliance item is a significant non-conformity or deficiency in the quality system that affects or has the potential to affect the quality of the product. A typical major non-compliance is the exclusion of one of the ISO clauses. Sometimes many third-party auditors will issue a

---

TABLE 4.3
Audit Corrective Action Report (ACAR) for Internal Audit

Facility:	ACAR:	Date Audited:
Lead Auditor:	Document Number:	
Auditor:	Guide:	
Discussed with:	Verified with:	
Statement of the Standard or Requirement:		
Observation:		
Additional Comments:		
Recommendations:		

TABLE 4.4
Audit Corrective Action Report (ACAR) for External Audit

Facility:	ACAR:	Date Audited:
Lead Auditor:	Document Number:	
Auditor:	Guide:	
Discussed with:	Verified with:	
Statement of the Standard or Requirement:		
Observation:		
Additional Comments:		
Recommendations:		

---

major NC if there are several minor NCs related to each other. A major NC quite often is enough to withheld the certification.

A minor NC item is an isolated non-conformity that does not represent a system deficiency. However, if a series of minor non-compliances are identified, that may constitute a major deficiency. It is of paramount importance that ALL (major and minor) gaps (deficiencies) be fixed because they are indeed NCs which means they are deficiencies in the system or are not fulfilling the customer's requirement(s).

An earlier example may again clarify this point. As an auditor, early in the audit you observe a handwritten change in a document control record without the specified signature for authorization. You make a note of it and report the non-conformity to the area supervisor who assures you that this event must be an oversight. You proceed with the audit and find no more problems with missing signatures or anything else. The lack of signature is indeed a minor NC. On the other hand, after finding the missing signature and receiving assurances that the event was an isolated case of plain oversight, you find that changes in control documents are performed without appropriate signatures or proper authorization throughout the organization. At that point, you should record the incident as a major non-compliance or NC. Now it has become a major issue.

The difference between the first and second situations is that in the first case the lapse was an isolated error, whereas the second case implies a breakdown of the system and as such a major non-compliance.

2. *Closing meeting*: The closing meeting is sometimes called the post-audit conference (exit) meeting. The closing meeting must take place regardless of how rushed all the parties are. The meeting itself is attended by all managers of the audited activities as well as by all auditors. It is not recommended that the only auditee representative be the quality assurance manager; he or she may filter or distort the information. The purposes of the closing meeting are to

- Review the audit scope.
- Review the audit limitations.
- Present the summary of results.
- Identify all action requests.
- Clarify details and emphasize findings.
- Announce the registration eligibility.
- Present the manuscript report.
- Explain follow-up and response time (if applicable). Announce the post registration (surveillance) requirements.

If there is a follow-up report, then the contents of the report may contain the following:

- Purpose and scope of the audit.
- Participants.
- Background information.
- Summary of results.
- Identified weaknesses and strengths.
- Responses to action requests.
- Evaluation of responses.
- Timetable for re-audit (if necessary).
- Disposition on the certification.

The report should state the items of concern very carefully. Some guidelines are:

- Use management terms whenever possible. Avoid general jargon.
- Use familiar industry jargon whenever possible.
- Put comments in order of importance.
- Report results in a language that is concise, exact, and easy to understand.
- Call attention to truly exemplary practice whenever possible.
- The audit report should avoid emotional words and phrases, personal bias, nitpicking, and management overload.

So, what is the result (product) of phase 3? In this phase, the audit team produces a physical report that includes the findings of the audit and the recommendation whether or not the auditee should be certified.

Phase 4 – Closure

All audits come to an official close with phase 4. A typical closure may be a review of the non-compliances and the auditee's reaction to those non-compliances, or it may be a revisit to the organization by the Registrar. In the first case, a desk audit (review) may be necessary, whereas the second case may require a more formal approach, such as a physical audit. The closure of any audit focuses on four specific areas:

1. *Evaluation of the non-compliances and the action requested*: As a general rule, it is the responsibility of the lead auditor to evaluate the auditee's responses. The evaluation deals with
   - Evaluating and closing all corrected items.
   - Re-auditing the items that need on-site verification/validation.
   - Accepting or rejecting the corrections.
   - Assessing the chance of success if partial acceptance is required because of time limitations or some other legitimate reason.
   - Evaluating prevention measures to avoid repeating of the problems found.
2. *Follow-up on the corrective action*: All non-compliances require action by the auditee. The corrective action is a formal way to recognize the problem and to solve it in a systematic way. There are four basic steps in any corrective action:
   - Identifying the root cause of the problem with tools such as brainstorming, cause-and-effect diagrams, statistical process control (SPC) charts, 3×5 Why analysis, and process flow diagrams.
   - Identify the "escape point." That is where the problem originated, but it was not caught.
   - Specifying actions to correct the problem in the short term.
   - Specifying actions to correct the root cause in such a way that the problem will not repeat (preventive actions). This is the solution for the long term.
   - Identifying responsibilities and the timetable for corrective action. The lead auditor is responsible for evaluating both the timeliness and the effectiveness of the resources applied to the corrective actions. Most Registrars require a 30–60 days (45 days average) buffer zone for the corrective actions for major non-compliances.
3. *Retention of audit documents*: Finally, the closure in any audit must consider the retention of both official and unofficial records. Official records are generally retained for 3–5 years. However, in some industries, the retention period may be defined from cradle to grave, as for example in the nuclear industry. The following items are official audit records:
   - Notification letter.
   - Audit plan (if separate document).
   - Preassessment questionnaire.
   - Checklist.
   - On-site audit/evaluation.

- Auditee's response.
- Closure letter.
- Auditors qualifications (optional).

    Unofficial records are retained for 6–12 months. Unofficial audit records include the auditor's workpapers, supporting documents, and any other additional correspondence.

4. *Communication with management*: Part of the responsibility of the lead auditor is to make sure that management gets the message (the results of the audit) and incorporates these results into the corrective action identified in the quality system. This is a prescription of the ISO 9001 clause 7.4. It is expected that improvements will occur through corrective action initiatives and that the system will become more effective through a preventive strategy.

SURVEILLANCE

To receive certification from a Registrar is only the beginning commitment to a quality system that needs monitoring to ensure continued compliance with the standard. The actual monitoring varies from Registrar to Registrar; however, all Registrars have some kind of monitoring system. Some rely on regular, unannounced audits; others reassess at regular intervals (3 years is common) with one minor audit every 6 months.

Since one of the objectives of the certification is to *assure confidence* in the system, the idea of the surveillance is to make sure that the assessed organization continues to maintain its quality program. Atypical surveillance may cover the following:

- Maintenance of the internal audit program and appropriate corrective action.
- Customer complaints and their follow-up.
- Satisfactory completion of all corrective actions agreed upon at the previous audit (internal or third party).
- Appropriate use of the Registrar's logo.
- Follow-up on all NC items.
- Aspects of the quality system, possibly guided by the recorded non-conformities and minor non-conformities from the last audit.
- Whether top management utilizes internal audits for continual improvement purposes.

COMMON PROBLEMS IN CONDUCTING AUDITS

Just like any other endeavor, audits are subject to mistakes, problems, and misunderstandings. Some of the most basic and avoidable problems are as follows:

- *Inadequate planning and preparation for the audit*: The more you plan and prepare, the better the chances for a successful audit. You ran never plan or prepare enough.

- *Lack of clearly defined scope and objective*: You must know the boundaries and plan the objectives of what you are trying to accomplish. Take time to evaluate alternate proposals.
- *Inadequate procedures*: It has been reported (Irwin Professional Publishing, 1993; QIM, 1993) that one of the most frequent inadequacies for certification is lack of documentation. Without proper documentation, quality awareness, and the contribution of the operator, it is impossible to have a complete, accurate, and current quality system documentation. For the mandatory audit items, see Chapter 5.
- *Lack of properly trained auditors*: Without appropriate knowledge on how to perform the audit, the results of the audit may be worthless. It takes special skill and knowledge to perform an audit correctly. Investment in training is a must.
- *Lack of operator involvement*: Without the cooperation of all the employees in any organization, the audit will not be effective and will not fulfill the spirit of the ISO and any other requirement. Without their cooperation, procedures, and instruction, the information gathered may not be current and/or correct.
- *Lack of follow-up*: Without systematic follow-up, the audit and its results will become just one more activity in the organization. Active follow-up must be cultivated and encouraged at all times. Appropriate feedback must be given and corrective actions must be communicated throughout the organization on a timely basis.

A GUIDE TO THE GUIDE

The person charged with the responsibility to guide the auditors in the auditee's facilities is the guide. This is a very important responsibility and the success or failure of the audit may depend on the way the guide carries out the assigned tasks. The guide must understand both the mechanical aspects and the behavioral aspects of the audit.

1. *The Mechanical aspects*: These are the logistic considerations of the audit. The guide must be familiar with physical area where the audit will occur and should know how to get there by the shortest and safest path. The guide is also expected to know (a) the standard in which the company is being certified, (b) details of the organization's quality system, and (c) management policy as it relates to the quality system and operating procedures.
2. *The Behavioral aspects*: The guide's responsibility is to accompany the auditors at all times and answer their questions when asked. If a guide does not know the answer, he or she
   - Should not give an opinion.
   - Should not admit not knowing the answer.
   - Should know the source where the answer may be found and direct the auditor to that source.

The guide is the expert in the particular area; however, one of the guide's responsibilities is to help the auditors and the organization perform a successful audit. As such, a guide should try to accommodate the auditor and help the organization at the same time. Examples of how this policy may prove beneficial follow:

- If the auditor is about to ask a specific question of an operator and the guide knows that the operator does not have the answer, the guide should advise the auditor of the fact (and in some cases offer an explanation).
- A guide is not authorized or qualified to disagree with the auditor when the auditor identifies a non-compliance. However, based on his or her knowledge of both the standard and the process, you may diplomatically question a non-compliance if the guide suspects that the auditor does not understand the specific process or that the auditor is applying the wrong standard to the particular situation. If the non-compliance is not changed, the guide should report to the ISO coordinator or the department management. Under no circumstances should the guide argue with the auditor.
- The guide should sign the non-compliance or the ACAR even if he or she disagrees with the finding, for *the guide is a witness* and not an approving entity for the NC. The auditor is merely doing his or her job. The auditor is asking you to be a witness of the finding, not whether you agree that the finding is indeed a non-compliance. (Sometimes if the guide is not available, the operator is asked to sign as a witness of the finding.)
- The guide should always answer the questions of the auditor in a matter of fact and never editorialize or give a personal view and certainly the guide should never assume.
- The guide's cooperation with the auditor will save time and facilitate a productive audit. The organization requested the audit, and the objective is not only to be certified against a specific standard but also to improve the quality system.
- The guide's job is to help the auditor; the auditor is responsible for carrying out the audit.
- Finally, it is the auditor's job to find non-compliances; it is not the guide's job to supply them.

COMMUNICATIONS

All auditors talk, observe, and listen; in fact, one may say that the act of auditing is a matter of communication. The job of all auditors is to collect information based on objective evidence. However, the better the auditor is at communicating (both verbally and nonverbally), the more effective the audit will be. There are at least three ways auditors can communicate, and each method has its own advantages. The three ways are verbal (one or two way), nonverbal (mannerisms and body language), and observation:

1. *Verbal*: In a one-way communication, the auditor issues notices, procedures, letters, etc. When the auditee responds to the item, then the communication becomes two way. The auditor may prefer one-way communication because it is fast and because it gives the originator a measure of protection and feeling of security, as the communicated message is not questioned. Two-way communication is slower, less orderly, and certainly noisier. In two-way communication, the receivers will understand the message better and will feel more motivated to actively participate because they can check their understanding of what is required.

In the course of the audit, the auditor must **always** be alert to the communication rules that will make or break the audit. Some of the "always" rules are as follows:
- Be cognizant of your own abilities.
- Seek to be understood and to understand. Be a good listener.
- Simplify and clarify the message. Use the KISS (keep it simple Stephanie) principle.
- Make sure that the message has been understood.
- Examine the purpose of the message.
- Remember that the message may have long-term effects.
- Consider the human and physical setting.
- Make sure that your actions suit the message.
- Ask *to be sure*, before communicating. (If you are not sure, ask for clarification before you proceed to the next item.)

In case some of the communication rules seem well beyond the duty of an auditor, let us examine the issue of communication. People communicate to get something done, affect somebody's attitude, and reinforce the message. To accomplish these three goals, the auditor must
- Know the audience and be sensitive to age, race, sex, ethnicity, etc.
- Know the specific situation and why he or she is present.
- Know the message and organize the message in a way that communicates to the entire audience. The message should be informative rather than demeaning. It should address the experience level and knowledge of the audience.

2. *Nonverbal*: An auditor must be cognizant of all unusual changes in mannerism and body language in the course of the audit. There is no guarantee that these changes are meaningful, but a thoughtful follow-up question may prove invaluable in uncovering problem areas.

3. *Observation*: Observation is the major skill in any auditor's profile, the "bread and butter" of the trade. It takes much discipline in listening and communicating thoughts and actions to do a good job in observation. The auditor is asked to record what he or she observes without any bias. Without good observations, there cannot be a good audit. The following items describe some common communication mistakes.

Incorrect	Correct
Assuming be	Complete and factual.
Asking any question	Ask open-ended questions with specific focus.
Acting as though you are the expert behavior	Be flexible; be enthusiastic; reinforce appropriate behavior and clarify your role
Using Jargon	Use everyday language; paraphrase; give feedback frequently; and especially when asked.
Jumping around	Be systematic and logical; be well organized; be positive; point out possibilities.
Being very formal	Be personable and friendly; be willing to discuss issues.
Using incorrect grammardiction, punctuation, or pronunciation	Learn to use the language effectively, make the language work for you.
Using pronouns and other words incorrectly	Learn when to use *it is* (it's) as opposed to *its*; learn the difference between datum and data. For a thorough review of proper language, use any writer's handbook, a thesaurus, or even a dictionary.

COST OF QUALITY AND LIABILITY

Two of the most misunderstood concepts of the standards and the auditing process are *cost of quality* and *liability*. Neither of these topics are part of certifiable standards (ISO 9001). However, they are either explicitly and/or implicitly mentioned in the ISO 9004 (clauses 4.2; 5.2; 7.1.2; 8.4.2; 9.5.3; 9.7 (c)).

One of the questions auditees usually ask is how the evaluation of cost and liability is going to take place. The answer, of course, is very simple. Although the certifiable standards do not provide a direct link between cost, liability, and quality, the ISO 9004 identifies such a requirement.

A good auditor can identify both these issues by going around to other clauses of the certifiable clauses for hints and directions. For example, in the case of cost of quality, an auditor may get to the same information in ISO 9001 through clauses 8.5 (product and service provision), 8.7 (control of non-conforming outputs), and elements 9 (performance evaluation) and 10 (improvement). In the case of liability, an auditor may get to similar or the same information through clauses 8.3.4 (design and development controls) and 8.5.2 (identification and traceability).

INTERNAL VERSUS EXTERNAL AUDITORS

As we mentioned earlier, the internal auditor (usually not certified by an outside body) conducts the audits for his or her organization, and the results of the audit are not recognized for certification. The internal auditor more often than not is an employee of the auditee. The external auditor is usually certified by an outside body and conducts the audit as an impartial entity. The results of this kind of audit, if run under the jurisdiction of a Registrar, may be recognized for certification. An external auditor is never an employee of the auditee.

CERTIFICATION VERSUS NONCERTIFICATION OF AUDITORS

Whether you are certified in the art of auditing is an academic point. What is essential is that you demonstrate competence and specific knowledge of what you are about to audit. If you are interested in becoming a professional auditor, then the requirement of certification is mandatory. If, on the other hand, you want to be an auditor for your organization, then certification is not necessary. The ISO 9001 clause 9.2.2 states that internal audits must be performed; it does not specify that the auditors must be certified. The moral of this clause is that knowledge of the organization and the product line and auditing skills are indeed requirements for a good internal audit. If that is all that is expected, then certification is a moot point.

To attain certification in auditing is a personal goal and a worthy pursuit. You must attend a recognized course and complete the requisites. For a current requirement list contact the ASQ. Nevertheless, certification is not a requirement for internal audits or for unofficial audits to your supplier base.

Many commercial programs in auditing are available. Their only requirement is that you pay the seminar fee and attend the usual 2-day class recommended by the RAB or some other certifying body. (We may add that the content of the certifiable course and the noncertifiable course is practically the same. The major differences are the higher fees, the test, the limited number of locations, and the limited class size (usually a maxi number of 20 students) for the certifiable course. A certifiable class will qualify you to perform official audits if you pass the examination at the end of the course – provided you also have the appropriate experience and the other requisites for the certification).

ETHICS

Ethics is the discipline that deals with what is good and what is bad, with moral duty and obligations. Additionally, ethics may be defined as those principles of conduct that govern an individual or a group.

From the audit perspective, ethics involves a built-in moral component. This component, in conjunction with the self-correcting mechanisms of the audit (the objective and systematic evidence), guards the moral integrity of the auditors. Without this moral integrity, auditors can interpret, conduct, or declare to be true things that are not so. Moral integrity is precisely what will keep the system and the auditors honest and correct. Ethical principles of auditing practices; the activity of auditing (both the process and priority); the treatment of the auditee; and the allocation of proper, appropriate, and correct credit for discoveries are most essential in any audit. The quality audit is no exception.

Additional ethical principles, which are often unrecognized, should guide the analysis of the audit data. Let me explain. I was assigned to be the lead auditor in an ISO 9001 audit in Houston, Texas. Just after the opening meeting and before the actual audit, one of the auditors approached me and asked, "Dean, can I bother you for a moment?" I thought, of course, that the auditor had a question regarding the audit itself, the standard, or something relating to the audit. To my surprise, the auditor, with a most serious expression, asked, "I have worked with many lead auditors, and every one of them has guidelines for the first audit. What I want to know is how many non-compliances you want me to deliver for this audit?" I disqualified him on the spot.

I hope the example illustrates the subtlety of the auditor's question. He was not going to be objective in the audit. I am sure he would have delivered whatever number of non-compliances I identified. He would not have conducted an impartial audit, and his results (data driven as they may have been) would have been useless.

A few more examples of poor auditing practices may delineate the point of ethics and understanding of audits. The first example is from training – a single-group pretest–posttest design. Between the pretest and the posttest, a treatment (the content of the course) was administered to all subjects, a treatment that was expected to increase the scores of the dependent variable. Suppose 4 of the 25 subjects obtained lower scores on the posttest than on the pretest. In other words, the treatment did not have a positive effect on all the subjects. These lower scores were removed from the data, and a correlated mean t-test was calculated based on the remaining 21 score pairs. The table probability associated with the obtained value of the statistic was less than 0.05, and the auditor, or the trainer in this case, concluded that the treatment was effective.

The second example is from a supplier survey. The auditor or the quality assurance engineer analyzes the responses from a supplier survey. The analysis led to the calculation of over 100 chi-square tests of independence. Of these, 20 were statistically significant at $p < 0.05$. In the report of the analysis, the auditor reported only the significant tests, making no mention of the more than 80 chi-square tests that did not yield significant values of the statistic.

Few auditors would fail to recognize the poor practices illustrated in these examples. However, it is very important to add that in neither example was the auditor intentionally deceptive. Furthermore, either auditor would be both surprised and righteously offended at insinuations of fraudulent practices. With these examples, we have tried to illustrate the lack of consideration of ethical principles in the analysis of the data and the subtle nature of our tendency as auditors to deceive ourselves about the objective evidence. Always remember that "figures do not lie; liars figure."

In the course of an audit – especially 1 that runs 5 or more days – there is abundant opportunity to test the ethics of even the best auditors. There are always minor shadings, manipulations, and distortions as well as fraudulent practices. However, auditors must protect the auditee whenever the evidence is not sufficient or questionable – to the point of always giving them the benefit of the doubt and always trying to be consistent, fair, and objective in evaluations. The excuse that *everybody distorts things just a little bit* and *everybody knows* it is totally unacceptable. We must do everything in our power to have appropriately trained auditors who know the standards, their intent, the methodology of the audit, and the wisdom to seek both advice and criticism about the audits they perform.

One of the author's area of ethical concerns is the evaluation of data, especially in the chemical or pharmaceutical industries or other industries that deal in batches. The issue of batch process is quite common, so from an auditor's point of view the questions are when, how, where, and what to sample. As it turns out, the answers to these very simple questions can be very troublesome for both the auditor and the auditee if they are not familiar with random and representative samples. Tukey (1980) suggested four general processes in which the researcher (auditor) influences the data:

1. The generation of appropriate questions. This is where a good checklist may prove worthwhile.
2. The process of research (audit) design. This is where the planning of the audit will prove priceless.
3. The monitoring of data collection. This is where the data has to be challenged as to its appropriateness and applicability.
4. The process of data analysis. This is where, from the author's experience, most of the inconsistencies and erroneous results happen. In fact, since this area is of profound interest to both auditor and the auditee, let us examine it further.

The American Statistical Association (ASA) developed ethical guidelines for statistical practice (Ad Hoc Committee on Professional Ethics, 1983). Five elements of these guidelines are directly pertinent to the ethics of data analysis. First, the results and findings should be presented honestly and openly. Surprises and contradictory findings must be avoided. Second, deceptive or untrue statements in the reports should be avoided. Third, the boundaries of inference should always be clear. Fourth, clear and complete documentation of data editing, statistical procedures, and assumptions must be provided. Fifth, all statistical procedures should be applied without concern for a favorable outcome.

Against the backdrop of the ethical guidelines suggested by the ASA, seven categories of questionable practices are described below. They should not be interpreted as a complete list of ethical problems but rather as a sample of commonly encountered practices.

1. Selectivity of data (cooking the data) or altering or discarding data that appears to be anomalous. The idea comes from Dunnette (1966) who proposes that auditors tend to support their hypotheses. Very famous cases of *cooking the data* have been reported by Birch (1990), Westfall (1973), McNemar (1960), and others.
2. Use of data-driven hypotheses. This is when an auditor may look at patterns in the data and then decide what hypotheses to test.
3. Use of postmortem analyses. This technique of data analysis guarantees specific findings based on subset analysis in the quest for significance. This can be interpreted as nitpicking or the auditor is after you.
4. Probability pyramiding and selective reporting. This technique guarantees results by analyzing the data with every analysis known. Sooner or later one will identify significance. Diaconis (1985) and Nehrer (1967) have reported this problem of multiplicity.
5. Type I and Type II calculations bias. Friedlander (1964) pointed out this subtle bias resulting from the interaction between calculation errors and

hypotheses. Specifically, our commitment to our results leads us to recheck our calculations and/or observations only in some circumstances and to avoid such rechecking in others. When the original calculations of statistics do not support our hypotheses, we will recheck the computations. However, when the initial analysis of the data supports our hypotheses, we fail to see the need for such a check and potential errors are undetected.

6. Confusion of probability level and the strength of relationships. Here the results will be interpreted the same way for both the small and large samples whether there is a relationship or not. Examples are given by Dunnette (1966), Thompson (1988), and McNemar (1960).

7. Confusion between *exploratory* and *confirmatory* approaches to analysis. Tukey (1980) and Brush (1974) have addressed this issue quite extensively. In summary, this issue has to do with the formation of the hypothesis in the exploratory study or analysis. However, regardless of what the study or analysis shows, the hypothesis must be subjected to independent, confirmatory inquiry.

One may wonder at this point why an auditor would have a bias, and if the bias exists, where did it come from? This is a fair question but not so easy to answer. As auditors we take pride in what we do – no question about it. We are professionals, and we are eager to do the best possible job. However, in the process of doing our best, we become like tunnel-vision parents who can see no fault with our offspring. We convince ourselves that the empirical data would speak in favor of our theories if only we could analyze them properly or longer. From this perspective, our audit efforts become directed towards supporting our theories rather than testing them. (A good example of this is when auditors audit their pet areas, say document control or design or procurement, with extra interest and vigor. Their excuse may be that these areas are more demanding or complicated or more time consuming, but the fact is, that they like to spend time auditing their favored area).

Another area of creeping bias is the pressure – whether real or imaginary – to find non-compliances. Publications and consultants alike announce the likelihood of trouble spots of the audit on the first, second, or whatever try, and in addition, they point out alleged trouble-spot areas. As a consequence, auditors feel that if they pass the auditee on the first try, something went wrong, that they did not do their job. In addition, if they did not find anything wrong in the highlighted areas of possible errors or non-compliances, their professionalism is challenged.

This undue emphasis on statistical performance for the organization, the auditor, and the Registrar begins with the introductory training in the ISO standards. Even though the training and statistics are supposed to guide and direct, they take on an aura of their own and in the end work against both the auditor and the auditee.

A third possibility of ethical problems is the way the auditor defines success of audit activities. Some auditors want to be known as hard, or difficult, or easy, or likable, and they stop at nothing to convince everyone of their perception. They interpret hard or difficult as fair and unbiased. How unfortunate. These auditors have missed the essence of the ISO, and other standards which focus on substantiating the system and also on the corrective action and continual improvement of the organization. Audits should follow the rules of randomness, consistency, and fairness and not be concerned with the popularity of the auditor. Any audit should NOT (ever) be considered – under any circumstances – an inspection process.

RECOMMENDATIONS FOR IMPROVING ETHICAL PRACTICES

As we already have mentioned, auditors are professional individuals who are asked to evaluate objective evidence and offer a decision based on their findings. In the process of evaluation, the results are sometimes clouded for both individuals and organizations to the point where hard feelings and conflicts occur. To avoid hard feelings and conflicts, some recommendations are the following:

- Focus on honesty throughout the audit.
- Develop an awareness of the subtlety of ethical issues in data analysis.
- Train auditors to audit not to nitpick or spy.
- Develop checklists to guide you during the audit, but do not use checklists as tools to dwarf initiative, inquisitiveness, and creativity.
- Develop openness during the audit and evaluation of results.
- Avoid any situation in which the auditor was previously employed by the auditee, regardless of the reason for separation.
- Avoid any situation in which the auditor was previously employed by the auditee's competitor, regardless of the reason for separation.
- Avoid any situation in which the auditor has holdings of stocks or bonds in the auditee's business or that of a competitor of the auditee.
- Avoid any situation in which the auditor was associated with the development of the quality system under evaluation to meet a particular quality standard.

Auditors following guidelines such as the above will enter the audit arena with a strong commitment to fairness, an aversion to bias, and a high level of personal satisfaction. They indeed will do the best to evaluate the auditee.

CONFORMANCE VERSUS COMPLIANCE

Up to now we have used the terms conformance and compliance either by themselves or as a separate entity. However, because of their importance, it is important that both the auditors and auditees know the difference. When one talks about conformance, the implication is that there is a specification "somewhere" that defines what the customer "needs" and quite often "expects." As such, when ALL those needs and expectations are met, then conformance is reached. If not, there is a "gap," and the only way to close that gap is to fix the discrepancy.

So, when we talk about "certification," we mean that through confirmation of the system audited we found that all requirements have been met and the product or service meets the criteria of the customer.

When conformance is not met, the audited organization runs the chance of losing business – unless the NC(s) are corrected based on the expectations of the customer.

On the other hand, compliance is something that is a government regulation (Law) or an Occupational Hazard and Safety Regulation. A compliance is not a negotiable item. It must be fixed. The responsibility of fixing these non-compliances is primarily the task of management's leadership and implementation practices. What does this mean? Some examples are as follows:

- Get top-down commitment to full legal compliance.
- Initiate a program for risk assessment. Guide the appropriate individuals to a sound checklist for compliance.
- Initiate a benchmarking study with the "best" risk practices in the business.
- Investigate the "need" for training for any specializing safety or occupational hazards and provide that training to the individuals who are involved – as needed.
- Make sure that the appropriate, applicable responsibilities and scheduling are allocated on a timely manner.
- Be proactive on compliance incidents without delay.
- Make sure that there is a system of reporting non-compliance issues without any form of retaliation.
- Do not be intimidated with regular compliance audits.

When compliance is not met, the organization's management may pay a heavy fine and or go to jail. In addition, the organization may be given a probation period to fix the non-compliance.

VERIFICATION VERSUS VALIDATION

We have used both terms here as synonyms because of their general use in the process of auditing. However, the reader must understand that there are significant differences between the two terms, especially when they refer to problem-solving methodologies such as in FMEAs. In both of these instances, *verification is thought of as the planning activities* to contain and or fix the problem. *Validation is the process of making sure that your verification plan(s) will in fact fix the problem.*

THE ROLE OF THE CONSULTANT IN THE DOCUMENTATION AND PREASSESSMENT STAGES

The field of quality is becoming very complicated and at the same time very demanding. There are questions regarding the hiring of consultants. However, two questions are essential:

1. Do we really need a consultant? Before answering this question, let me remind the reader of the story about Lewis and Clark. Lewis and Clark were considered to be the best surveyors in the land. They certainly knew how to paddle and which way was west. That, however, did not stop them from hiring Sacagawea, a Native American guide familiar with the territory and the dangers that lurked there.

Having giving you a clue to my answer to this question, I must admit that strictly speaking, you do not need a consultant. Your organization can do it all. However, if you do it alone, it is guaranteed that you will repeat mistakes and that you will try things based on intuition and hearsay. Eventually you may succeed. On the other hand, if your organization wants to expand, become more efficient and save a tremendous amount of resources as part of the implementation and auditing process, then a consultant is the answer. A good consultant can be a coach, a changing agent, and leader.

2. How much is it going to cost? Just like anything else, the cost depends on how you negotiate and what services the consultant renders. Before your organization decides to hire a consultant, do the following:

a. *Shop around*: Find out what is available and ask questions. Ask for references. Find out about specific experiences and how the consultants will be able to help you.

b. *Be careful of promises and quick solutions offered by many consultants*: Remember, the ISO/IATF total implementation takes time. It is not an overnight success. Be realistic of what you are asking the consultant to do.

c. *Be choosy*: When your organization is ready to make the decision, pick what is the best for your organization. The best can be defined as being comfortable with the relationship you are about to embark on. Money is not always the best criterion for evaluation. Availability, experience, ability to get along with different people, good communication skills, and above all, a track record are all important and worth considering as part of your selection process. DO NOT select ONLY on PRICE.

So, how much is it going to cost? No single algorithm governs the way companies choose consultants. The best we can offer is to recognize that cost, effort, and time are interrelated. They all depend on *what*, *where*, and *why* the organizations buy. Remember, however, that quality can and in fact is being bought on a daily basis, from boutiques as well as department stores, and the customer is just as satisfied with the boutique purchase as with that from the department store. Let us all remember the words of W. D. Deming (2000): "Do not buy on price; buy on total cost."

5 Mandatory Auditing Items

All along we have been talking and referencing documents and records needed for the auditor to review for an effective audit. In this chapter, we have grouped them together for easier access for each standard individually (see special note).

AN OVERVIEW OF THE MANDATORY DOCUMENTS AND RECORDS REQUIRED BY ISO 9001:2015

Here are the documents you need to produce if you want to be compliant with ISO 9001:2015. (Please note that some of the documents will not be mandatory if the company does not perform relevant processes):

- Scope of the quality management system (QMS) (clause 4.3)
- Quality policy (clause 5.2)
- Quality objectives (clause 6.2)
- Criteria for evaluation and selection of suppliers (clause 8.4.1).

And, here are the mandatory records (note that records marked in *italics* are only mandatory in cases when the relevant clause is not excluded):

- *Monitoring and measuring equipment calibration records* (clause 7.1.5.1)
- Records of training, skills, experience, and qualifications (clause 7.2)
- Product/service requirements review records (clause 8.2.3.2)
- *Record about design and development outputs review* (clause 8.3.2)
- *Records about design and development inputs* (clause 8.3.3)
- *Records of design and development controls* (clause 8.3.4)
- *Records of design and development outputs* (clause 8.3.5)
- *Design and development changes records* (clause 8.3.6)
- Characteristics of product to be produced and service to be provided (clause 8.5.1)
- Records about customer property (clause 8.5.3)
- Production/service provision change control records (clause 8.5.6)
- Record of conformity of product/service with acceptance criteria (clause 8.6)
- Record of non-conforming outputs (clause 8.7.2)
- Monitoring and measurement results (clause 9.1.1)
- Internal audit program (clause 9.2)
- Results of internal audits (clause 9.2)
- Results of the management review (clause 9.3)
- Results of corrective actions (clause 10.1)

Non-mandatory Documents

There are numerous non-mandatory documents that can be used for ISO 9001 implementation. However, I find these non-mandatory documents to be most commonly used:

- Procedure for determining context of the organization and interested parties (clauses 4.1 and 4.2)
- Procedure for addressing risks and opportunities (clause 6.1)
- Procedure for competence, training, and awareness (clauses 7.1.2, 7.2, and 7.3)
- Procedure for equipment maintenance and measuring equipment (clause 7.1.5)
- Procedure for document and record control (clause 7.5)
- Sales procedure (clause 8.2)
- Procedure for design and development (clause 8.3)
- Procedure for production and service provision (clause 8.5)
- Warehousing procedure (clause 8.5.4)
- Procedure for management of non-conformities and corrective actions (clauses 8.7 and 10.2)
- Procedure for monitoring customer satisfaction (clause 9.1.2)
- Procedure for internal audit (clause 9.2)
- Procedure for management review (clause 9.3)

THE DETAILED MANDATORY DOCUMENTS REQUIRED BY ISO 9001:2015 ARE

1. Documented information to the extent necessary to have confidence that the processes are being carried out as planned (clause 4.4).
2. Evidence of fitness for the purpose of monitoring and measuring resources (clause 7.1.5.1).
3. Evidence of the basis used for calibration of the monitoring and measurement resources (when no international or national standards exist) (clause 7.1.5.2).
4. Evidence of competence of person(s) doing work under the control of the organization that affects the performance and effectiveness of the QMS (clause 7.2).
5. Results of the review and new requirements for the products and services (clause 8.2.3).
6. Records needed to demonstrate that design and development requirements have been met (clause 8.3.2).
7. Records on design and development inputs (clause 8.3.3).
8. Records of the activities of design and development controls (clause 8.3.4).
9. Records of design and development outputs (clause 8.3.5).
10. Design and development changes, including the results of the review and the authorization of the changes and necessary actions (clause 8.3.6).

11. Records of the evaluation, selection, monitoring of performance, and re-evaluation of external providers and any actions arising from these activities (clause 8.4.1).
12. Evidence of the unique identification of the outputs when traceability is a requirement (clause 8.5.2).
13. Records of the property of the customer or external provider that is lost, damaged, or otherwise found to be unsuitable for use of its communication to the owner (clause 8.5.3).
14. Results of the review of changes for production or service provision, the persons authorizing the change, and necessary actions taken (clause 8.5.6).
15. Records of the authorized release of products and services for delivery to the customer including acceptance criteria and traceability to the authorizing person(s) (clause 8.6).
16. Records of non-conformities, the actions taken, concessions obtained, and the identification of the authority deciding the action in respect of the non-conformity (clause 8.7).
17. Results of the evaluation of the performance and the effectiveness of the QMS (clause 9.1.1).
18. Evidence of the implementation of the audit program and the audit results (clause 9.2.2).
19. Evidence of the results of management reviews (clause 9.3.3).
20. Evidence of the nature of the non-conformities and any subsequent actions taken (clause 10.2.2).
21. Results of any corrective action (clause 10.2.2).

These documents may look like a long list of busy work, as they need to be retained as records of the results of your QMS. However, there is a silver lining for those who think this form of documentation is excessive and burdensome. The good news is that many organizations already have these records as they exist in their current practice of their operation. What needs to be done – in some cases – is the reconfiguration of the existing documentation to align with the new requirements. Newer businesses (or those new to the ISO 9001:2015 standard) who don't have a long or broad documentation history are the ones who usually spend the most time generating the paperwork listed above.

Ultimately, the documented information is part of the core value of the ISO 9001:2015. It encourages you to standardize the processes you already employ and to work towards consistent data collection and data updates to core paperwork like the documents listed above.

A GUIDE TO THE NON-MANDATORY DOCUMENTATION

In addition to the list of mandatory items, an organization as we mention earlier, may choose to add documents that contribute to the value of the organization's QMS. These documents are called *non-mandatory* and they include – but they are not limited to the following list:

- Organization charts
- Procedures
- Specifications
- Work instructions
- Test instructions
- Production schedules
- Approved supplier lists
- Quality plans
- Strategic plans
- Quality manuals
- Internal communication documents

If one does create these types of documented information, then one must follow the same rules laid out in clause 7.5. In other words, treat them the same way as you treat the named required documented information.

THE DETAILED MANDATORY DOCUMENTS REQUIRED BY ISO 14001:2015 ARE

- Scope of the environmental management systems (EMS) (clause 4.3)
- Environmental policy (clause 5.2)
- Risk and opportunities to be addressed and processes needed (clause 6.1.1)
- Criteria for the evaluation of significant environmental aspects (clause 6.1.2)
- Environmental aspects with associated environmental impacts (clause 6.1.2)
- Significant environmental aspects (clause 6.1.2)
- Environmental objectives and plans for achieving them (clause 6.2)
- Operational control (clause 8.1)
- Emergency preparedness and response (clause 8.2)

And, here are the mandatory records:

- Compliance obligations record (clause 6.1.3)
- Records of training, skills, experience, and qualifications (clause 7.2)
- Evidence of communication (clause 7.4)
- Monitoring and measurement results (clause 9.1.1)
- Internal audit program (clause 9.2)
- Results of internal audits (clause 9.2)
- Results of the management review (clause 9.3)
- Results of corrective actions (clause 10.1)

NON-MANDATORY DOCUMENTS

There are numerous non-mandatory documents that can be used for ISO 14001 implementation. However, the following non-mandatory documents are the most utilized:

- Procedure for determining the context of the organization and interested parties (clauses 4.1 and 4.2)
- Procedure for identification and evaluation of environmental aspects and risks (clauses 6.1.1 and 6.1.2)
- Competence, training, and awareness procedure (clauses 7.2 and 7.3)
- Procedure for communication (clause 7.4)
- Procedure for document and record control (clause 7.5)
- Procedure for internal audit (clause 9.2)
- Procedure for management review (clause 9.3)
- Procedure for management of nonconformities and corrective actions (clause 10.2)

THE DETAILED MANDATORY DOCUMENTS REQUIRED BY OH&S 18001 ARE

- Scope of the OH&S management system; 4.1, 4.4.4
- OH&S policy; 4.2, 4.4.4
- OH&S objectives and programme(s); 4.3.3, 4.4.4
- Roles, responsibilities, and authorities; 4.4.1
- Communication from external parties; 4.4.3.1
- OH&S management system elements and their interaction; 4.4.4
- Operational control procedures; 4.4.6

MANDATORY RECORDS

- Competence, awareness, and training records; 4.4.2
- Record of hazard identification; 4.3.1
- Risk assessment, significance, and controls; 4.3.1
- Performance monitoring records; 4.5.1
- Calibration records; 4.5.1
- Evaluation of compliance records; 4.5.2.1, 4.5.2.2
- Non-conformity, corrective, and preventive action records; 4.5.3
- Internal audit records; 4.5.5
- Management review records; 4.6

These are the documents and records that are *required* to be maintained for the OHSAS 18001 management system. However, you should also maintain any other records that you have identified as necessary to ensure your management system can function, be maintained, and improve over time.

ADDITIONAL DOCUMENTATION

However, all "real-world" systems require more than these minimum mandatory requirements to ensure a robust and reliable system. There are several additional documented procedures which are commonly employed to ensure a robust and reliable system, including:

- Procedure for hazard identification, risk assessment, and determining controls; 4.3.1
- Procedure for non-conformity, corrective action, and preventive action; 4.5.3
- Procedure for monitoring and measurement; 4.5.1
- Procedure for legal and other compliance requirements; 4.3.2
- Procedure for competence, training, and awareness; 4.4.2
- Procedure for evaluation of compliance; 4.5.2
- Procedure for control of documents and records; 4.4.5, 4.5.4
- Procedure for internal audit; 4.5.5
- Procedure for emergency preparedness and response; 4.4.7
- Procedure for communication, participation, and consultation; 4.4.3

In the final analysis, to determine what additional documentation you require, simply ask yourself the question; *do we need a documented procedure to ensure consistency between employees?* Therefore, a good rule to follow is that simple, clear, documentation is more effective than complicated documentation in ensuring that all employees can deliver repeatable outcomes. And don't forget, procedures can be written, in the form of a flowchart, use pictures etc. – choose the most effective!

THE DETAILED MANDATORY DOCUMENTS REQUIRED BY ISO 45001:2015 ARE

- Scope of the OH&S management system (clause 4.3)
- OH&S policy (clause 5.2)
- Responsibilities and authorities within OH&SMS (clause 5.3)
- OH&S process for addressing risks and opportunities (clause 6.1.1)
- Methodology and criteria for assessment of OH&S risks (clause 6.1.2.2)
- OH&S objectives and plans for achieving them (clause 6.2.2)
- Emergency preparedness and response process (clause 8.2)

MANDATORY RECORDS

- OH&S risks and opportunities and actions for addressing them (clause 6.1.1)
- Legal and other requirements (clause 6.1.3)
- Evidence of competence (clause 7.2)
- Evidence of communications (clause 7.4.1)
- Plans for responding to potential emergency situations (clause 8.2)
- Results on monitoring, measurements, analysis, and performance evaluation (clause 9.1.1)
- Maintenance, calibration, or verification of monitoring equipment (clause 9.1.1)
- Compliance evaluation results (clause 9.1.2)

- Internal audit program (clause 9.2.2)
- Internal audit report (clause 9.2.2)
- Results of management review (clause 9.3)
- Nature of incidents or non-conformities and any subsequent action taken (clause 10.2)
- Results of any action and corrective action, including their effectiveness (clause 10.2)
- Evidence of the results of continual improvement (clause 10.3)

NON-MANDATORY DOCUMENTS (PROCEDURES)

There are numerous non-mandatory documents that can be used for ISO 45001 implementation. The following non-mandatory procedures are the most commonly used:

- Procedure for Determining Context of the Organization and Interested Parties (clause 4.1)
- OH&S Manual (clause 4)
- Procedure for Consultation and Participation of Workers (clause 5.4)
- Procedure for Hazard Identification and Assessment (clause 6.1.2.1)
- Procedure for Identification of Legal Requirements (clause 6.1.3)
- Procedure for Communication (clause 7.4.1)
- Procedure for Document and Record Control (clause 7.5)
- Procedure for Operational Planning and Control (clause 8.1)
- Procedure for Change Management (clause 8.1.3)
- Procedure for Monitoring, Measuring, and Analysis (clause 9.1.1)
- Procedure for Compliance Evaluation (clause 9.1.2)
- Procedure for Internal Audit (clause 9.2)
- Procedure for Management Review (clause 9.3)
- Procedure for Incident Investigation (clause 10.1)
- Procedure for Management of Non-conformities and Corrective Actions (clause 10.1)
- Procedure for Continual Improvement (clause 10.3)

THE DETAILED MANDATORY DOCUMENTS AND RECORDS REQUIRED BY IATF 16949: 2016 ARE

1. Scope of the QMS (clause 4.3)
2. Documented process for the management of product safety-related products and manufacturing processes (clause 4.4.1.2)
3. Quality policy (clause 5.2)
4. Responsibilities and authorities to ensure that customer requirements are met (clause 5.3.1)
5. Results of risk analysis (clause 6.1.2.1)
6. Preventive action record (clause 6.1.2.2)

7. Contingency plan (clause 6.1.2.3)
8. Quality objectives (clause 6.2)
9. Records of customer acceptance of alternative measurement methods (clause 7.1.5.1.1)
10. Documented process for managing calibration/verification records (clause 7.1.5.2.1)
11. Maintenance and calibration record (clause 7.1.5.2.1)
12. Documented process for identification of training needs including awareness and achieving awareness (clause 7.2.1)
13. Documented process to verify competence of internal auditors (clause 7.2.3)
14. List of qualified internal auditors (clause 7.2.3)
15. Documented information on trainer's competency (clause 7.2.3)
16. Documented information on employee's awareness (clause 7.3.1)
17. Documented process to motivate employees (clause 7.3.2)
18. Quality manual (clause 7.5.1.1)
19. Record retention policy (clause 7.5.3.2.1)
20. Documented process for review, distribution, and implementation of customer engineering standards/specifications (clause 7.5.3.2.2)
21. Registry of customer complaints (clause 8.2)
22. Product/service requirements review records (clause 8.2.3.2)
23. Procedure for design and development (clause 8.3.1.1)
24. Record about design and development outputs review (clause 8.3.2)
25. Documented information on software development capability self-assessment (clause 8.3.2.3)
26. Records about product design and development inputs (clause 8.3.3.1)
27. Records about manufacturing process design input requirements (clause 8.3.3.2)
28. Document a process to identify special characteristics (clause 8.3.3.3)
29. Records of design and development controls (clause 8.3.4)
30. Documented product approval (clause 8.3.4.4)
31. Records of design and development outputs (clause 8.3.5)
32. Manufacturing process design output (clause 8.3.5.2)
33. Design and development changes records (clause 8.3.6)
34. Documented approval or waiver of the customer regarding the changes in design (clause 8.3.6.1)
35. Documented revision level of software and hardware as part of the change record (clause 8.3.6.1)
36. Documented supplier selection process (clause 8.4.1.2)
37. Documented process to identify and control externally provided processes, products, and services (clause 8.4.2.1)
38. Documented process to ensure compliance with statutory and regulatory requirements of purchased processes, products, and services (clause 8.4.2.2)
39. Documented process and criteria for supplier evaluation (clause 8.4.2.4)
40. Records of second-party audit reports (clause 8.4.2.4.1)
41. Characteristics of product to be produced and service to be provided (clause 8.5.1)

42. Control plan (8.5.1.1)
43. Total productive maintenance (TPM) system (clause 8.5.1.5)
44. Records of traceability (clause 8.5.2.1)
45. Records about customer property (clause 8.5.3)
46. Production/service provision change control records (clause 8.5.6)
47. Documented process to control and react to changes in product realization (clause 8.5.6.1)
48. Documented approval by the customer prior to implementation of the change (clause 8.5.6.1)
49. Documented process for management of the use of alternate control methods (clause 8.5.6.1.1)
50. Record of conformity of product/service with acceptance criteria (clause 8.6)
51. Record of expiration date or quantity authorized under concession (clause 8.7.1.1)
52. Documented process for rework confirmation (clause 8.7.1.4)
53. Record on disposition of reworked product (clause 8.7.1.4)
54. Documented process for repair confirmation (clause 8.7.1.5)
55. Record of customer authorization for concession of the product to be repaired (clause 8.7.1.5)
56. Notification to the customer about the non-conformity (clause 8.7.1.6)
57. Documented process for disposition of non-conforming product (clause 8.7.1.7)
58. Record of non-conforming outputs (clause 8.7.2)
59. Monitoring and measurement results (clause 9.1.1)
60. Internal audit program (clause 9.2)
61. Results of internal audits (clause 9.2)
62. Documented internal audit process (clause 9.2.2.1)
63. Results of the management review (clause 9.3)
64. Action plan when customer performance targets are not met (clause 9.3.3.1)
65. Results of corrective actions (clause 10.1)
66. Documented process for problem solving (clause 10.2.3)
67. Documented process to determine the use of error-proofing methodologies (clause 10.2.4)
68. Documented process for continual improvement (clause 10.3.1)

In addition to the specific requirements mentioned in the standards and/or the organization's specifications, the auditor must carefully verify, validate, and evaluate the existence and effectiveness of the following items:

- *Maintenance*
    - A planning process that achieves reliability excellence, with manufacturing, operations, and maintenance working together. Comprehensive tasks are developed and delivered that support TPM.
    - The spare parts and their storage are managed.
    - The critical parts are identified.

- The maintenance planning must cover machines and tools for preventive maintenance (PM) and where applicable the PM should exist and if not, it should be developed.
- *Training*
  - Training plan and timetable exists for each employee aligned to the business plan based upon job requirements and evaluations.
  - Training process is standardized and effective. Flexibility chart is updated for all operation.
- *Supply chain management*
  - Tier supplier targets are defined and their performance are tracked. Annual audits are performed and issues found (these may not have developed into problems yet) are tracked until closed.
  - Quality data is used in the sourcing decision process.
- *FMEAs*
  - Have – as applicable – a design FMEA.
  - All operations must have been analyzed for risk using a process FMEA. PFMEA training or workshops must be done by cross-functional teams, including manufacturing team member input. Risk should be prioritized by (a) severity, (b) criticality (severity × occurrence), and (c) Risk Priority Number (RPN) – severity × occurrence × detection. If the RPN is used (not recommended) make sure the numerical values are consistently applied.
  - Failure modes are included/comprehended in the PFMEA (*i.e.* wrong parts, mixed parts).
  - Make sure that the failures are not effects or causes; if in doubt, ask the question: *what can go wrong with the function*? There are at least six basic failures: (a) no function, (b) intermittent function, (c) unintendent (surprise) function, (d) partial function, (e) wear out – degradation – function, and (f) over function.
- *Capability*
  - Capability reviews of process equipment with high risk or impact (based on the key characteristic designation system) are held to identify process capability (>1.33 P_{pk}) and appropriate and effective corrective actions. A reaction plan for non-capable process is present. Corrective actions are documented so that stability is monitored and used for process capability.
- *Capacity*
  - Capacity review and stream allocation analysis based on 5- and 6-day production rates with an increase in margin of 10% minimum.
  - Review the Overall Equipment Effectiveness (OEE) – aim for at least 85%.
- *Visual standards and controls*
  - Review the applicability of the visual factory.
  - Review the usage of the 5S.
  - Review for consistency. ALL visual standards are common within the facility (*e.g.* multiple lines) as well as between facilities building

the same platform or product line – this applies for global facilities, if applicable.

- *Visual/tactile/audible standards – communicated and understood*
  - Visual standards – if appropriate and applicable – are clearly communicated to the team members (or the individual operators) at the work station and incorporated or referenced in the procedures and/or work instructions.
  - Team and/or operators have been trained to identify the visual standards with appropriate training.
  - Visual standards must be easily able to differentiate *good from bad* parts and certainly they must satisfy customer requirements. For a detailed discussion of visual standards see Ortiz and Park (2010).
- *Error proofing verification*
  - All error-proofing (detection) devices are checked for function (failure or simulated failure) at the beginning of the shift, if not, follow the process control plan (PCP).
  - If appropriate and applicable, the "Red Rabbits" parts are calibrated.
  - Error proofing masters or challenge parts (if, and when) used must be clearly identified.
  - Records of verification are available and reviewed as appropriate and applicable.
  - Reaction plan is standardized and understood in case of error-proofing device(s) malfunction.
- *Deviation management (bypass the standard and or requirements)*
  - The organization plant must identify manufacturing processes and error-proofing devices which can be placed in deviation or completely bypassed from normal operations.
  - ALL risks for all deviations and bypass items are evaluated and reviewed.
  - Standard work instructions are available for each deviation and/or bypass process.
  - Implemented bypass and deviations must be reviewed regularly as the goal is to reduce or eliminate any deviations and or bypasses.
- *Measurement system analysis (Gage R&R, gage calibration)*
  - Gage capability (Gage R&R, linearity, bias, stability, *etc.*) of monitoring and measuring equipment is determined and the equipment is certified/calibrated with traceable standards at a scheduled frequency.
- *Development of process control (PFMEA, PCP, SW)*
  - PFMEA, process control plans and standardized work (SW) documentation are comprehensive, sufficient, and flow one from the other.
  - Critical operations are identified with the appropriate symbol (*e.g.* ∇) at the operation and in the standardized work documentation (*e.g.* work instructions).
- *Process control plan implemented (PCP).*
  - PCP checks are performed at the correct frequency and sample size.
  - Sample size and frequency must be determined based on risk and occurrence of the cause and ensures that the frequency adequately protects

the customer by ensuring that the product represented by the inspection or test does not reach the customer before the results of these inspection or tests are known.

- Sample and frequency must be reviewed on regular basis or as appropriate.
- Checks are documented using the proper control method (*e.g.* SPC – control charts, check sheets.
- Reaction plan(s) from the PCP are present, followed, and effective.
- Effective PCPs must include prevention plan(s) in addition to corrective action(s).
- Process-specific requirements are met and audit records kept.
- Action plans with both corrective and prevention plans must be created to close the gap.
- *Process change control*
  - Plant processes are validated with a Gemba visit and the process flow diagram relative to changes in design and/or process (5M&E).
  - The plant follows a documented change control process for customers and internal changes.
  - The DFMEA and PFMEA must be updated as needed and reviewed at least once a year.
  - All changes (engineering and manufacturing and when appropriate, suppliers may participate) must be the result of meetings with the appropriate departments.
- *Change control – production trial run (PTR)*
  - A reasonable sample size is used in production trial run (PPAP – AIAG; PPAP – Phase 0, Ford Motor Co.) based on risk and confidence level. (Usually 300 samples.)
  - Parts are contained, stored, and clearly identified prior to and after the PTR and PPAP approval.
- *Layered audit prevention plan*
  - Layered audits are in place to assess compliance to standardized processes, identify opportunities for continuous improvement, and provide coaching opportunities.
  - Layered audit is owned by management. Audit plan must include multiple levels of management.
  - Audits are tracked and their results recorded.
  - All non-conformances must be followed up, corrected, and provide an acceptable resolution by the auditor(s) and the organization's management.
- *Process audit*
  - Audits are tracked and their results recorded.
  - All non-conformances must be followed up, corrected, and provide a prevention plan.
  - Using the SIPOC model evaluate each of the components and follow-up on non-conformities – if they exist. For a very detailed discussion, see VDA (2016).

- *Standardized work*
  - All work is documented using a standard format and meets all safety, quality, all standard and organizational requirements.
  - Workplace organization is implemented (*e.g.* 5S: 5S is a workplace organization method that uses a list of five Japanese words: *seiri* (整理), *seiton* (整頓), *seisō* (清掃), *seiketsu* (清潔), and *shitsuke* (躾). These have been translated as "Sort," "Set In order," "Shine," "Standardize," and "Sustain." The list describes how to organize a work space for efficiency and effectiveness by identifying and storing the items used, maintaining the area and items, and sustaining the new order. The decision-making process usually comes from a dialog about standardization, which builds understanding among employees of how they should do the work. In the United States during the last 10 years, the list has been amended with an extra S for safety (safe). So, now instead of 5S, it is known as 6S. For more detailed information, see Galsworth (2005) and Gapp et al. (2008).
  - Standardized work has to be detailed enough to ensure that operation is performed on standardized way on each cycle.
- *Rework/repair/confirmation/tear down*
  - Repairs (online and offline) are compliant with approved standardized work.
  - Repaired, reworked, or replaced material is processed at a minimum through an independent repair confirmation (second person or machine after repair).
  - Reintroduction of worked part includes all downstream checks in order to ensure that all control plan inspections and tests are performed.
- *Alarm and escalation*
  - Non-conforming product having sufficient alarm limits with escalation alarms are responded to according to the alarm and escalation process (reaction plan).
- *Non-conforming material and material identification*
  - Team members have standardized work and understand what to do with non-conforming items.
  - Conforming items are identified, handled, and stored appropriately.
  - Non-conforming and suspect material must be identified, quarantined, and evaluated for disposition.
  - An appropriate and applicable method of containment must be established that is effective and will ensure a bad product will not reach the customer. (Third-party inspection may be convenient and fast but is not the best way to containment.)
  - Traceability is applied according to acceptable and agreed-upon methods of the finished product and reworked parts when needed.
- *Team problem-solving process*
  - A well-developed standardized problem-solving process (*e.g.* 8D, 5-Why, Six Sigma: DMAIC, 3 × 5-Why) exists at all levels of the organization.

- Formal problem-solving activities are initiated according to complexity of the issue or problem and specific criteria. For detail discussion on different tools, see Tague (2005) and Anderson and Fagerhaug (2000).
- Issues are identified, root causes analyzed, and robust actions completed in a timely manner.
- Problem-solving is driven at the team level, and all teams are involved with a common goal, even though they are cross-functional and multi-discipline.
- Leaders are actively involved in coaching and guiding the process.
- *Andon system implementation*
  - A well function Andon system is implemented in all production areas to support the team members when abnormal conditions occur and communicate relevant information. (An Andon system is one of the principal elements of the Jidoka quality control method pioneered by Toyota as part of the Toyota Production System and therefore now part of the lean production approach (Everett and Sohal, 1991; Liker, 2004).) It gives the worker the ability, and moreover the empowerment, to stop production when a defect is found and immediately call for assistance. Common reasons for manual activation of the Andon are part shortage, defect created or found, tool malfunction, or the existence of a safety problem. Work is stopped until a solution has been found. The alerts may be logged to a database so that they can be studied as part of a continual improvement process. The system typically indicates where the alert was generated and may also provide a description of the trouble. Modern Andon systems can include text, graphics, or audio elements. Audio alerts may be done with coded tones, music with different tunes corresponding to the various alerts, or prerecorded verbal messages.
  - All operational areas of the organization are using the Andon process as intended and this shows tangible (measurable) results on the operating floor.
- *Inspection gates (verification, validation, and final inspection)*
  - Final inspection – of some sort – must be in place. It could be 100% if nothing else is available.
  - All items must be verified for checking and validated as appropriate.
  - All quality checks are included in standardized work.
  - Successive and more frequent checks may be required for complex or difficult problems.
- *Fast response process*
  - Exit criteria with appropriate timing and defined for closing issues.
  - There is an awareness of corrective actions for all applicable workers.
  - Fast response with predefined methodologies and/or tools (5D, health charts, tracking sheets, PFMEA, control plan, internal audit results, etc.).
  - As necessary and depending on the severity or magnitude of the problem meetings, reviews and the lie are mandatory and recorded.

- *Quality-focused checks*
  - High-risk items from critical operations have a quality-focused check performed each shift.
  - High-risk quality-focused items from customer feedback and problem-solving are included in internal audits or other pertinent sources.
- *Feedback/feed forward*
  - There is a feedback system. The system is appropriate and applicable to forward and backward information (verification and validation) from the final inspection station to manufacturing and vice versa.
  - Quality alerts are handled appropriately. Generally, they are issued by engineering for pending problems. They do have a time limit of approximately 90 days. If nothing happens during that window, then the alert reverts back to the original specification.
  - SREAs – Supplier Request for Engineering Approval must be initiated (generally) by suppliers (manufacturing). The turnaround is about 72 hours. A change is requested and engineering must approve it before it is implemented. If engineering does not approve it, the original design remains intact. It must be appropriately approved and recorded.
- *PFMEAs risk reduction and annual review*
  - Are the PFMEAs reviewed frequently to identify possible risk reduction?
  - Are the failures, effects, and root causes appropriately identified?
  - Are there more than one failure, effect, and root cause identified?
  - Are the FMEAs following a standardized format (AIAG, VDA, or customer-specific approach)?
  - Action plans for top issues must be identified.
  - Recommendations, responsibility, and closure time must be included.
- *Material-handling process and FIFO first in, first out*
  - A plant FIFO or material-handling process is documented and practiced in all operations.
  - Visual aids assist in process flow. They are used if appropriate and applicable.
- *Shipping approval packaging events*
  - Material is shipped in the designated production container with proper labeling for regular production and all saleable build.
- *Safety*
  - Appropriate documentation is necessary to demonstrate that the organization is focused on safety.
  - Systems are in place to reduce safety risks.
  - Communication processes are in place to assure that safety concerns are discussed by all appropriate personnel.
  - A safety system exists to identify concerns and are being tracked, reviewed, and addressed on a timely basis.
- *Contamination requirements*
  - Contaminated product is secure and quarantined.
- *Inspection gates* (verification and validation stations including final inspection)

Special note: Within an external audit rarely there have been situations where lead auditors and auditors want to view documented evidence of organizational context, stakeholder needs, and risks. This request is very unusual, because nowhere in the standards and/or requirements are mentioned as either required or not. Therefore, whether a supplier submits to this request, it is up to the organization. Of course, the availability should be determined based on the risk that such a forthcoming gesture can create in the organization of doing so or not (Fonseca and Domingues, 2017).

6 Acronyms

Generally, acronym lists appear at the front of the book. Here we have made the decision to make it a separate chapter for two reasons: (1) this list is quite lengthy and (2) the automotive acronym list is unusual and somewhat more detailed than others. The acronym list is not something unique to the automotive industry. All organizations, all standards, and all specifications have their own language (jargon) to facilitate communication. That jargon, especially in the automotive industry, is somewhat more complicated because many of the acronyms have a different meaning even though they are spelled the same. For any auditor doing a desk or a distance audit, this may present a difficult task to decipher the contents of the documentation presented to them. Therefore, in this list – although not an exhaustive one – we hope that it covers the most frequent used in the automotive industry and it will provide a quick reference for those who get stumbled.

4D: A problem-solving methodology focusing on: Discovery, Design, Development, and Delivery.

5D: A five-step approach to problem solving. A shortcut to the Global 8D.

8D: Even though it is called 8D, in reality its official name is G8D which is a nine-step methodology for problem solving. This methodology was developed by Ford Motor Company in the early 1980s, and after several iterations of improvement, currently it has become one of the most common problem-solving technics, not only within Ford, but also in the entire industry. In fact, many other industries are using it with excellent results. It is based on nine steps, which are D0 = Prepare for the 8D process, D1 = Establish the team, D2 = Describe the problem, D3 = Develop Interim containment action (ICA), D4 = Define and verify root cause and escape point, D5 = Choose and verify permanent corrective actions (PCAs), D6 = Implement and validate permanent corrective actions, D7 = Prevent recurrence, and D8 = Congratulate the team. The difference between 8D and G8D is the additional step (D0) added to the G8D.

14D: An elaborate 14-step problem-solving methodology

3K: Kiken (dangerous), Kitanai (dirty), Kitsui (stressful) – general workplace hazards

3M: Muda (waste), Mura (irregular, inconsistent), and Muri (unreasonable strain)

5M: Manpower, machine, method, material, and measurement – sources of variation

5M&E: Manpower, machine, method, material, and measurement and environment – sources of variation

3P: Production, preparation, process

3R: Recording, recalling, reconstructing – when generating new ideas

3S: Stabilize, synchronize, standardize – steps in lean product development

4P: Production, process, prove-out, program

5P: Plant, production, people, policies, procedures – sources of variation (for fishbone)

7P: Proper prior planning prevents pitifully poor performance

5R: Responsiveness, reliability, rhythm, responsibility, relevance

5W2H: Who, what, when, where, why and how and how many (root cause analysis)

4S: Arrangement (sort), organization (Seiton), cleanliness (Seiso), act of cleaning (Seiketsu)

5S: Seiri (sort), Seiton (straighten), Seiso (shine), Seiketsu (standardize), and Shitsuke (sustain)

6S: Seiri (sort), Seiton (straighten), Seiso (shine), Seiketsu (standardize), and Shitsuke (sustain), safety

7 Wastes (Sins): Overproduction, transport, waiting, inventory, defects, over-processing, unnecessary movement

AAA:	American Automobile Association
ABS:	Affordable Business Structure
AIAG:	Automotive Industry Action Group
AIC:	Accelerated Implementation Center
AIM:	Automated Issues Matrix
AIMS:	Automated Issues Matrix System
AME:	Advanced Manufacturing Engineering
AMPPE:	Advanced Manufacturing Pre-Program Engineering
ANOVA:	Analysis of Variance
AP:	Attribute Prototype
APEAL:	Automotive Performance Execution and Layout
APQP:	Advanced Product Quality Planning
ASQ:	American Society for Quality
AV:	Appraiser Variation
AVT:	Advanced Vehicle Technology
AWS:	Analytical Warranty System
AXOD:	Automatic Transaxle Overdrive Transportation
B&A:	Body and Assembly Operations (replaced by Vehicle Operations)
BCG:	Business Consumer Group
BIC:	Best in Class
BIQS:	Built-In Quality Supply
BIS:	Body Shop Information System
BLI:	Business Leadership Initiative
BOM:	Bill of Materials
BTB:	Bumper-to-Bumper
BTS:	Build to Schedule
BUR:	Business until Review
CA:	Customer Attribute
CA:	Corrective Action
CAD:	Computer-Aided Design

CAE:	Computer-Aided Engineering
CAP:	Corrective Action Plan
CAR:	Capacity Analysis Report
CAR:	Corrective Analysis Report
CAS:	Capacity Analysis Sheet
CBG:	Consumer Business Group
CC:	Critical Characteristic
CC:	Carbon Copy
CC:	Change Cut-Off
CC:	Courtesy Copy
CCC:	Costumer Concern Classification
CDS:	Component Design Specification
C/E:	Cause and Effect
CET:	Campaignable Events Team
CETP:	Corporate Engineering Test Procedures
CETs:	Common External Tariffs
CFR:	Constant Failure Rate
CIM:	Computer-Integrated Manufacturing
CIWG:	Continuous Improvement Work Group
CL:	Centerline
CMM:	Coordinate Measuring Machine
CMMS:	Common Material Management System
CMMS3:	Common Manufacturing Management System 3
Code X:	Pre-build focusing on exterior components
Code Y:	Pre-build focusing on interior components
CP:	Common Position
CP:	Confirmation Prototype
Cp:	Relates the allowable spread of the specification limits to the measure of the actual variation of the process
CPE:	Chief Program Engineer
Cpk:	Measures the process variation with respect to the allowable specification and takes into account the location of the process average
CPU:	Cost per Unit
CQDC:	Corporate Quality Development Center
CQIS:	Common Quality Indicator System
CR:	Concern Response
CRS:	Concern Resolution Specialist
CRT:	Component Review Team
CSA:	Corporate Security Administrator
CSI:	Customer Service Index
DCO:	Duty Cycle Output
DCP:	Dynamic Control Plan
DDL:	Direct Data Link
Df:	Degrees of Freedom. Sometime, denoted as *df*.
DFA:	Design for Assembly
DFM:	Design for Manufacturability

DFMEA: Design Failure Mode and Effects Analysis
DFR: Decreasing Failure Rate
DFR: Design for Reliability
DMA: Database Maintenance Administrator
DOE: Design of Experiment
DOM: Dealer Operations Manager
DP: Design Parameters
DQR: Durability Quality and Reliability
DTD: Design to Delivery
DTD: Dock to Dock
DV: Design Verification
DVM: Design Verification Method
DVP: Design Verification Plan
DVP&PV: Design Verification Process and Production
DVP&R: Design Verification Plan and Report
DVPR: Design Verification and Product Reliability
DVPV: Design Verification and Process Verification
EAO: European Automotive Operations
EASI: Engineering and Supply Information
ECAR: Electronic Connector Acceptability Rating
EDI: Electric Data Interchange
EESE: Electrical and Electronic System Engineering
EMM: Expanded Memory Manager
EMS: Environmental Management System
EOL: End of Line
EQI: Extraordinary Quality Initiative
ES: Engineering Specifications
ESI: Early Supplier Improvement
ESP: Extended Service Plan
ESTA: Early Sourcing Target Agreement
ESWP: Early Sourcing Work Plan
EV: Equipment Variation
F&T: Facility and Tooling
FACT: Facilitation and Certification Training
FAO: Ford Automotive Operations
FAP: Ford Automotive Procedure
FAQ: Frequently Asked Questions
FASS: Field Action/Stop Shipment
FA/SS: Field Action/Stop Shipment (this is the preferred acronym)
FCPA: Ford Consumer Product Audit
FCSD: Ford Customer Service Division
FDVS: Ford Design Verification System
FER: Final Engineering Review
FER: Fresh Eyes Review
FEU: Field Evaluation Union
FIFO: First In First Out

FMEA:	Failure Mode and Effect Analysis
FMVSS:	Federal Motor Vehicle Safety Standards
FOB:	Ford of Brain
FPDS:	Ford Production Development System
FPS:	Ford Production System
FPSI:	Ford Production System Institute
FPS IT:	Ford Production System Information Technology
FQRs:	Frequent Quality Rejects
FR:	Functional Requirements
FRG:	FAO Reliability Guide
FS:	Final Sign-Off
FS:	Final Status
FSIC:	Ford System Integration Control
FSN:	Ford Supplier Network
FSS:	Full Service Suppliers
FTEP:	Ford Technical Educational Program
FTT:	First Time Through
FUNC-APPRV:	Functional Approvals
FVEP:	Finished Vehicle Evaluation Program
GAP:	Global Architecture Process
GC:	Global Craftmanship
GCARS:	Global Craftsmanship Attribute Rating System
GCEQ:	Global Core Engineering Quality
GEM:	Generic Electronic Module
GIS:	Global Information Standards
Global 8D:	Global 8D – A nine-step problem-solving methodology
GPIRS:	Global Prototype Inventory Requisition and Scheduling
GPP:	Global Parts Pricing
GQRS:	Global Quality Research System
GRC:	Government Regulations Coordinator
GRC:	UN-ECE Group des Raporteous de Ceintures
GR&R:	Gage Repeatability and Reproducibility
GRVW:	Gross Vehicle Weight
GSDB:	Global Supplier Data Base
GSSM:	Global Sourcing Stakeholders Meeting
GYR:	Green, Yellow, Red
HI:	High Impact
HIP:	High Impact Characteristic (Priority Characteristic)
HR:	Human Resources
HTFB:	Hard Tool Functional Build
HVAC:	Hating Ventilating and Air-Conditioning
IATF-16949:	Industrial Automotive Standard
ICA:	Interim Containment Action
ICCD:	Intensified Customer Concern Database
IE:	Industrial Engineer
IFR:	Increasing Failure Rate

ILVS:	In-Line Vehicle Sequencing
IM:	Industrial Materials
IP:	Instrument Panel
IPD:	In Plant Date
IQ:	Incoming Quality
IQS:	Initial Quality Study
IR:	Internal Reject
ISO:	International Organization for Standardization. It is NOT an acronym. It comes from the Greek word *iso* which means equal
ISO-9000:	International Standard for Quality. The basic standard for all organizations
ISPC:	In-Station Process Controls
J1:	Job One
JIT:	Just in Time
JPH:	Jobs per Hour
JSA:	Job Safety Analysis
KKK:	PSW not Ready for Approval and/or Inspection
KLT:	Key Life Testing
KO:	Kick Off
LCL:	Lower Control Limit
LDEM:	Lean Design Evaluation Matrix
LOA:	Letter of Agreement
LP&T:	Launch Planning and Training
LR:	Launch Readiness
LRR:	Launch Readiness Review
LS:	Launch Sign-Off
LSL:	Lower Specification Limit
LTDB:	Light Truck Data Base
MBJ1:	Months before Job One
MBO:	Manufacturing Business Office
ME:	Manufacturing Engineering
MIS:	Months in Service
MMSA:	Material Management System Assessment
MOD:	Module
MP&L:	Material Planning and Logistics
MPPS:	Manufacturing Process Planning System
MRB:	Material Review Board
MRD:	Material Required Date
MS:	Material Specifications
MS3 (MSIII):	Material Supply Version III
MTC:	Manage the Change
MY:	Model Year
NAAO:	North American Automotive Operations
NFCASR:	No Fiat-Chrysler Automotive Customer-Specific Requirements
NFCSR:	No Ford Customer-Specific Requirements
NFM:	Noise Factor Management

NGMCSR:	No General Motors Customer-Specific Requirements
NIST:	National Institute of Standards and Testing
NMPDC:	New Model Program Development Center
Nova C:	New Overall Vehicle Audit
NTEI:	New Tooled End Items
NVH:	Noise, Vibration, Harshness
OCM:	Operating Committee Meeting
OEE:	Overall Equipment Effectiveness
OEM:	Original Equipment Manufacturer
ONP:	Owner Notification Program
OS:	Operational System
OS:	Operator Safety
OTG:	Open to Go
PA:	Program Approval
PAG:	Premier Automotive Group
PAL:	Project Attribute Leadership
PAT:	Program Action Team
PAT:	Program Activity Group
PAT:	Program Attribute Team
PCA:	Permanent Corrective Action
PCI:	Product Change Information
PD:	Product Development
P Diagrams:	Parameter Diagrams
PDL:	Product Design Language
PD Q1:	Product Development Q1
PDSA:	Plan, Do, Study (replaced the Check phase), Act. Known as the Deming Cycle
PFMEA:	Process Failure Mode and Effect Analysis
PH:	Proportions and Hardpoints
PI:	Process Improvement
PIPC:	Percentage of P_{pk} Indices Process Capable
PIST:	Percentage of Inspection Point That Satisfy Tolerance
PM:	Preventive Maintenance
PM:	Program Manager
PM:	Project Management
PMA:	Project Management Analyst
PMT:	Project Management Team
PMT:	Project Module Team
PO:	Purchase Order
POC:	Point of Contact
POT:	Process Ownership Team
Pp:	Process Potential
PPC:	Product Planning Committee
PP&T:	Product Planning and Technology
PPAP:	Production Part Approval Process
Ppk:	Process Capability

PPL:	Program Parts List
PPM:	Parts per Million (applied to supplier's defective parts)
PPPM:	Program and Pre-Production Management
PR:	Product Readiness Milestone
PR:	Product Requirement
PR:	Public Relations
PS1:	Pre SI Milestone 1
PS2:	Pre SI Milestone 2
PSO:	Production Standard Order
PSS:	Private Switching Service
PST:	Program Steering Team
PSW:	Part Submission Warrant
PT:	P/T Design Complete
P/T:	Power Train
PTO:	Power Train Operations
PTR:	Platinum Resistance Thermometer
PV:	Part Variation
PV:	Process Variables
PV:	Production Validation
PVBR:	Prototype Vehicle Build Requirements
PVM:	Production Validation Method
PVP:	Powertrain Validation Program
PVT:	Plant Vehicle Team
PVT:	Product Vehicle Team
QA:	Quality Assurance
QC:	Quality Control
QCT:	Quality Cost Timing
QFD:	Quality Function Deployment
QFTF:	Quality-Focused Test Fleet
QLS:	Quality Leadership System
QMS:	Quality Management System
QOE:	Quality of Event
QOS:	Quality Operational System
QPM:	Quality Program Manager
QPS:	Quality Process System
QR:	Quality Reject
QSA-PD:	Quality System Assessment for Product Development
QTM:	Quality Team Member
QVA:	Quality-Focused Value Analysis Workshop
R:	Range
R/1000:	Repairs per Thousand (one of the man reliability/warranty metrics)
RAP:	Remote Anti-Theft Personality Module
REDPEPR:	Robust Engineering Design Process Enabler Project
RIE:	Reliability Improvement Engineer
R&M:	Reliability and Maintainability
RMS:	Resource Management System

ROA:	Return on Assets
ROCOF:	Rate of Occurrence of Failure
RPN:	Risk Priority Number
RRCL:	Reliability and Robustness Check List
RRDM:	Reliability and Robustness Demonstration Matrix
RRR:	PSW Rejected
R&R:	Repeatability and Reproducibility
R&R:	Roles and Responsibilities
R&VT:	Research and Vehicle Technology
RWUP:	Real World Usage Profile
s:	Standard Deviation for Sample. The Standard Deviation for population is the Greek letter σ
S^2:	Variance for Sample. The variance for population is the Greek letter σ^2
SC:	Significant Characteristic
SC:	Strategic Confirmation
SCAC:	Supplier Craftmanship Advisory Committee
SCs/CCs:	Significant Characteristics/Critical Characteristics
SCTs:	Strategic Commodity Teams
SDS:	Subsystem Design Specification
SDS:	System Design Specification
SEVA:	Systems Engineering Value Analysis
SHARP:	Safety and Health Assessment Review Process
SI:	Strategic Intent
SI:	System International des Unit
SIM:	Supplier Improvement Metrics
SMART:	Synchronous Material and Replenishment Trigger
SME:	Subject Matter Expert
SMF:	Synchronous Material flow
SOW:	Statement of Work
SP:	Strategic Planning
SP/AP:	Structural Prototype/Attribute Prototype
SPC:	Special Product Committee
SPC:	Statistical Process Control
SP&PI:	Strategic Process and Product Improvement
SPROM:	Sample Promise Date
SREA:	Supplier Request for Engineering Approval
SRI:	Supplier Responsible Issues
SSI:	Sales Satisfaction Issues
SSM:	Strategic Sourcing Meeting
ST:	Surface Transfer
STA:	Supplier Technical Assistance
STARS:	Supplier Tracking and Reporting System
SVC:	Small Vehicle Center
TA:	Target Agreement
TAP:	Target Achievement Plan
TCM:	Total Cost Management

TED:	Things Engineers Do
TEG:	Tool and Equipment Group
TEM:	Total Equipment Management
TGR:	Things Gone Right
TGW:	Things Gone Wrong
TIS:	Time in Service
TOC:	Table of Contents
TPM:	Total Preventive Maintenance
TPPS:	Torque Process Potential
TQC:	Total Quality Control
TQC:	True Quality Characteristics
TRIZ (Russian):	Theory of Inventive Problem Solving
TRMC:	Timing, Release, and Material Control (also known as Tar-Mac)
TSP:	Technical Skill Program
TTO:	Tool Try-Out
UCL:	Lower Control Limit
USL:	Upper Specification Limit
VC:	Vehicle Center
V/C:	Very or Completely Satisfied
VC Buyer:	Vehicle Buyer Center (now Consumer Business Group Buyer)
VDI:	Vehicle Dependability Index
VDS:	Vehicle Descriptor Section
VDS:	Vehicle Design Specifications
VE:	Value Engineering
VER:	Vehicle Evaluation Rating(s)
VFG:	Vehicle Function Group
VIN:	Vehicle Identification Number
VLD:	Vehicle Line Director
VO:	Vehicle Office
VO:	Vehicle Operations
VOGO:	Vehicle Operations General Office
VP:	Vice President
VPMC:	Vehicle Project Management Coordinator
VPP:	Vehicle Program Plan
VQL:	Vehicle Quality Level
VQR:	Vehicle Quality Review
VRT:	Variability Reduction Team
VRT:	Vehicle Reduction Team
VRT:	Vehicle Review Team
WAS:	Work Analysis Sheet
WCR:	Worldwide Customer Requirements
WERS:	Worldwide Engineering Release System
WIP:	Waste in Process
WIP:	Work in Progress
WMI:	World Manufacturing Identifier
WPRC:	Warranty Parts Return Center
YS/YC:	Potential Significant and Critical Characteristics

7 Methodologies/Tools That the Auditor Should Be Familiar With

Any audit has its goal to verify and validate the system, product, and/or service that an organization has defined for itself. That verification and/or validation must be evaluated with objective evidence. Objective evidence is considered to be: multiple interviews with different individuals and documentation (standard, quality manual, procedures, instructions, records, and/or a combination of any of these).

That evaluation generally is conducted with either qualitative or quantitative data or a combination of both. In either case, the data evaluation may be complex with statistical analysis, heuristic or simple methodologies that have been used and proven successful over the years.

The approaches to the evaluation are many and impossible to cover them all here. However, an auditor is expected to be familiar with some basic and fundamental methodologies as well as individual tools to apply in the evaluation process. They are not expected to be experts in all tools, but their knowledge of them will enhance the credibility of any auditor and the audit results. So, here we will try to discuss in a rudimentary fashion some of the most common and effective ones.

VALUE STREAM MAPPING (VSM)

This methodology is fundamental for any auditor to understand because it gives an overview of the "process," and it helps in its analysis for improvement. A typical VSM is shown in Figure 7.1. For more information, the reader may want to consult a variety of quality books on the topic or a very good summary by Sonia Pearson at: https://tallyfy.com/value-stream-mapping/.

The inputs for a values stream map include all the resources you have and use to produce goods or services. The route you follow consists of information flow, value-adding steps, as well as their attendant non-value-adding steps. In a summary form and for the benefit of the auditor, a Value Stream Map (VSM) is a specific tool that allows you to see a top-down overview of the *total* business processes. Then, you can analyze the process or workflow, identifying wastes and inefficiencies. Typically, the following things are of general importance and should be reviewed:

- Delays that hold up the process
- Restraints that limit the process
- Excess inventory that ties up resources unproductively
- Overall Equipment Effectiveness (OEE)
- First-Time Throughput (FTT)

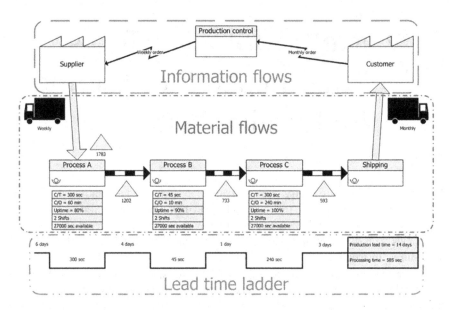

FIGURE 7.1   A typical VSM. Wikipedia, Daniel Penfield.

While value stream mapping is usually used for manufacturing processes, the same principles can apply to other industries too.

WHAT YOU NEED TO GET STARTED

If you are a second- or a third-party auditor, you will not be involved with the construction of a VSM. However, if you are an internal auditor, you more likely will be involved in the construction of it. So, if you are an internal auditor, first, the decision has to be made as to what needs to be mapped. In some businesses, one value stream map can cover just about everything the company does. This is especially true if your company produces a single product. On the other hand, if you have a complex mix of products or services, however, then you'd have to draw a separate map for each. With which process you'd start is, of course, up to you. Generally, though, you'd want to start off with the highest value or problem areas. As an auditor make sure you are looking at the specific auditing process that needs to be evaluated. Very rarely you will be interested in the VPM for the entire organization.

To actually carry out the mapping, you'd want to gather a small project team consisting of representatives from different departments. They have a first-hand perspective on how things are done and how well the current system works. You might even figure out several ways to improve the processes without even consulting the value stream map. Next up, you need a facilitator. This could be a senior manager who understands value stream mapping, or you can get an external consultant to help you. As you progress the work, you will create your map – but be ready to make changes as you go. Someone may just remember a missed step somewhere along the line, and that can change the whole picture. To actually draw the map, you can use:

- Pen and Paper – the simplest and preferred solution. Avoid the computerized version for at least the first or even the second draft.
- Flowchart Software – dedicated tools used for all sorts of business process mapping.
- Workflow Management Software – if applicable and available. In most cases, they are custom solutions for managing company workflows. In addition to simple mapping capabilities, you can also keep track of and manage the workflows.

VALUE STREAM MAPPING SYMBOLS

Symbols help with your visual overview. They are a short-hand version of communicating what the process is doing. They show exactly what kind of step you are dealing with. While you could always come up with your own symbols, it's usually easier to find an already established style and stick with it. There are many symbol tables in the internet – free of charge. The symbols are usually pretty intuitive – a simple line drawing of a pair of spectacles, *e.g.* indicates that someone has to "go and see," while a truck indicates transport, and so on.

SEVEN STEPS TO VALUE STREAM MAPPING

Now that you know the basics of value stream mapping, here are the exact steps you'd need to take to carry out the initiative of describing and evaluating.

Step #1: Decide how far you want to go (level of detail): Typically, you would start your mapping by indicating a start and end point. This would show where your internal process begins and ends. Some companies, however, prefer to map out the entire value chain. This, of course, has its pros and cons – while it does give you a better idea of the whole process, there's usually not much you can do about any external processes. Your average value stream map begins with the delivery of materials from direct suppliers and ends with delivery to the customer. Place the icon you have chosen to represent your starting and ending points on the left and right of your map. If your production processes are complex, you might decide to map each of the value-adding processes in greater detail after completing your overall map. In this case, you would start with the process that allocates the work as "supplier" and the process that receives it as the "client."

Step #2: Define the steps: Now determine what processes are involved in order to get from point A to point B. As a simple example, a nursery producing ornamental plants begins with seed from a supplier and delivers plants to a customer. Intervening steps that add value along the way might include the following:
  - Sowing
  - Transplanting
  - Growing

- Grading
- Shipping.

The more intermediate steps one comes up with, the more complexity is introduced in the process.

Step #3: *Indicate the information flows*: One of the advantages of value stream mapping is that it includes information flows. To continue with the example above, our plant nursery needs to place orders for its suppliers and its customers will place orders for delivery. How often is this done and how? Record it on your map. The teams or individuals responsible for each process that takes the product from input to output also need information. Where does it come from and how is this information passed on? Perhaps our flower grower has a centralized planning department which receives sales information and places orders with the seed supplier. It then uses this info and provides a weekly or monthly schedule for each of the processes. Add this department in the middle of the sheet between the input and output blocks, draw another block below it to indicate the weekly plan, and draw arrows from the plan to each of the departments it informs.

Step #4: *Gather the critical data*: You now have the basics, and it's time for an in-depth look at each process. To do so, you need real data and some of your mapping team might have to spend a little time collecting the information you need. Typical points to look at would include the following:

- The inventory items held for each process
- The cycle time (typically per unit)
- The transfer time (from step to step)
- The number of people needed to perform each step
- A number of products that must be scrapped
- The pack or pallet size that will be used
- The overall batch size that each step handles.

Step #5: *Add data and time lines to the map*: Once you have all the information, you can start adding it to your map. Draw a table or data box under each process block to do so. If you've used historical data, be sure to verify it using the current inputs and outputs for each process. Indicate the timeline involved in each process beneath your data blocks. This shows the lead time needed to produce products and the actual time spent on producing each unit, pack size, or batch. Don't be surprised if a product with a lead time of weeks takes just a few hours to produce. Make sure you report "real – actual" times and not the ones you would like to have. Historical or surrogate data are acceptable. However, they must be replaced with current data as soon as they become available.

Step #6: *Identify the seven wastes of lean:* Without a doubt, the VSM's purpose is to identify bottlenecks and eliminating them for an increase in productivity. So, just creating a value stream map without using it would be a complete waste of time. Now that you have one, it's time to start looking for the *seven wastes* that could be eating up your profits and restrict productivity. The classical wastes are as follows:

- **Transport** doesn't add value to your final product – unless you're in the transport business! See if you can reduce steps involving transport of materials or information that don't add value.
- **Inventory** of inputs and finished products costs you money which could have been earning income elsewhere. The lower your inventory levels can be without stonewalling production, the better it will be. (A good preventive way to look at the inventory is as a cushion for lack of production.)
- **Motion** costs time, and time is money. As an example, our nursery worker has to move her transplanted seedling 10 feet from the potting table to the tractor wagon. That's wasted time.
- **Waiting** because there's a bottleneck in a previous process or sub-process is another clear waste of valuable resources.
- **Over-processing** can be hard to gauge, but if an item can move from one process to another in an acceptable condition with less input, it should do so. Quite often this is cover up for rework.
- **Overproduction** is an additional pitfall to avoid. Even if your product isn't perishable, storing it and monitoring it until such time as a customer buys it is clearly a waste. (Overproduction is generally the genesis of future bottlenecks or in the management jargon sub-optimization of production. Unfortunately, this is a very common area of concern as many organizations use it to prove productivity measure. In doing so, quite often they sub-optimize their total production.)
- **Defects** mean reworking or scrapping and are clear money-eaters. Therefore, the focus for any defect consideration is to ask: How can you reduce defects in each step of the process you've mapped?

Step #7: Create the ideal (future) value stream map: You know how things are if you maintain the status quo, but how would you like them to look? Use your team to help you map out an ideal value stream map that eliminates, or at least reduces, all the wastes you spotted when analyzing the results of your value stream mapping exercise. Remind the team, often if you have to: *If you always do, what you always did, you will always get, what you always got.*

It's unlikely you'll be able to get there in one step, so you can create a series of intermediate future state maps. Your business would aim to reach these milestones at specific dates, and ultimately, they'd reach the goal you identified when you drew up your ideal state map.

What should you do now? Start the mapping process all over again! Few processes are so perfect that there's no more room for improvement! Your aim is nothing less than *operational excellence*.

Conclusion

The one drawback of value stream mapping done the old-fashioned way is the time that elapses between report-backs and meetings. For those hoping to slim down process flows fast, this can be frustrating. There's also the matter of monitoring the effects of changes you've decided to implement. Unforeseen, and possibly unwanted, consequences can flow from new parameters – or people could simply be getting it wrong because they aren't used to the new work method yet.

The best way to have that fix and speed things up is adopting the right technology, after the appropriate and applicable training has taken place for the employees that will be affected by the change(s).

Therefore, the idea for the auditor to be aware of the VSM of any organization is essential so that an *operational excellence* can be established and validated. However, what does operational excellence mean? In a simple definition it may be described as a **philosophy** that embraces problem-solving, prevention of problems, and leadership as the key to continuous improvement. People are often unsure of how to approach the subject of operational excellence. It is a difficult term to define, and most people either find the topic to be too ambiguous or too broad to talk about. However, it is **not a set of activities that you perform**. It's more of a **mindset** that should be present within the organization – at all levels. Now, you're probably thinking, "that sounds great in theory, but how do we implement this into actionable steps? Well, we're going to explain that in a bit. Before we get into implementing operational excellence, you need to understand how the concept is related to continuous improvement.

OPERATIONAL EXCELLENCE VERSUS CONTINUOUS IMPROVEMENT

Continuous improvement is the on-going effort to improve an organization's processes, products, or services. It usually takes place **incrementally over time** (the *Kaizen* approach), rather than instantly through some breakthrough innovation (quite often through a rigorous *reengineering*). By pursuing continuous improvement, an organization has a greater likelihood of continuing to maintain and build on these improvements. However, while continuous improvement is important, it is not enough on its own. As the organization continues to refine its process, product, or service, it needs **a way to continue to grow**. This is where operational excellence comes in. *Operational excellence* is a mindset that embraces certain principles and tools to create sustainable improvement within an organization. Or to put it more simply, operational excellence is achieved when every member of an organization is able to see the *flow of value to the customer*. Seeing it, however, isn't enough – they should actively try to improve *both the value, as well as its delivery*. Ultimately, operational excellence is not just about reducing costs or increasing productivity in the workplace. It's about creating the company culture that will allow you to produce valuable products and services for your customers and achieve *long-term sustainable growth*. Operational excellence is a journey that involves applying the right tools to the right processes. When this happens successfully, the ideal work culture is created where employees are provided for in a way that enables them to stay **empowered** and **motivated**.

Perhaps one of the very first steps that one must take for a successful Operational Excellence in the long term is taking the time to define the operational definition(s) at hand. An operational definition is the articulation of operationalization (what do we mean by the words we use to describe specific situations or tasks, or whatever is on the table). Furthermore, it is the way we define as a team the terms of a process (that includes, expected outcome, testing, validation as appropriate and so on) needed to determine the existence of an item or a specific (a variable, term, or object) and its properties (duration, quantity, extension in space, chemical composition, *etc.*). Since the degree of operationalization can vary itself, it can result in a more or less operational definition. The procedures included in definitions should be repeatable and accepted by anyone or at least by the stakeholders.

OPERATIONAL EXCELLENCE – SOME BASIC PRINCIPLES

Most of us working in the quality field are aware that every year the Shingo Institute of the Jon M. Huntsman School of Business gives out an award for operational excellence called the *Shingo Prize*. This prize is based on company culture, company results, and how well every employee demonstrates the Guiding Principles of the Shingo Model. Here is a summary of the principles based on https://shingo.org/ model/. Retrieved on May 10, 2020:

1. *Respect every individual:* The Shingo Model emphasizes that *everyone deserves respect,* because everyone has inherent worth and potential for success. However, it's not enough to have respect for others; you must demonstrate this respect to them as well. One of the best ways to demonstrate respect for your employees is by involving them in any necessary improvements to their department – especially if that change involves them. This will help them feel more empowered and motivated to contribute to the changes in a positive way. *To learn more about how to create a culture that engages every employee, from CEO to shop-floor staff, read the available literature on Kaizen.* (The classic book on the subject is: Imai, Masaaki (1986). *Kaizen: The Key to Japan's Competitive Success.* New York: Random House.)
2. *Lead with humility*: Leaders should always exercise humility. After all, the best improvements happen when people can acknowledge their shortcomings and look for a better solution. Humility involves a willingness to listen and take suggestions from everyone, regardless of that person's position or status within the company. Humility is to recognize your mistakes, apologize for undue behavior (in words and/or actions), and be ready to forgive. After all, we all make mistakes no matter what is our title and power status.
3. *Seek perfection*: This step in the model is often met with resistance as most people are quick to point out that perfection isn't possible. While perfection may feel unattainable that doesn't mean you can't *strive for it* anyway. By setting the bar high, you create a different mindset within your organization. When confronted with a problem, try to *look for long-term solutions* and always try to *simplify your work* without compromising the quality of the outcome. Try to implement the old adage of: aim for the stars and maybe you can go over the fence. If you aim for the fence, you may not get off the ground. This is not contrary to step # 2. Rather, it reemphasizes the notion

of perpetual endeavor to perfection, recognizing that we all make mistakes. If we do not try, we will never improve. We will not change.

4. *Embrace scientific thinking*: Innovation comes from *constant experimentation and learning*. Hence, it's always useful to know what works and what doesn't. By systematically exploring new ideas, you can encourage employees to do the same without fear of failure. By the way, quite often "a" failure may be the first step of the coming success.

5. *Focus on the process*: When things go wrong, there is a tendency to want to blame other people. In many cases, however, "the" problem is rooted in the process, not the person. This is because even *great employees* can't consistently produce ideal results with a *bad process*. When a mistake occurs, rather than immediately pointing fingers at the employees involved, assess what part of the process the error occurred in. Once you have done this, you can make adjustments to try to *achieve the expected results*.

6. *Assure quality at the source*: High quality can only be achieved once *every part of the process* is done correctly. It can be helpful to organize work areas in a way that will allow potential problems to become visible right away. When a mistake does occur, stop working immediately to correct the mistake before continuing. Making a defective part is not the best way to operate even though it may be reworked and sold anyway. By reworking it, we increase the cost of producing the part, and therefore, we are not receiving the anticipated profit from that item. In other words, we lose both productivity and profitability, by not emphasizing FTT. Rework may fix the problem and make it salable. However, the profit margin is NOT what the budget was for.

7. *Flow and pull value*: The objective of every organization is to provide maximum value to its customers. (Yes, an organization does have multiple customers – stakeholders.) Because of this, organizations should ensure that the process and workflow are continuous because interruptions create *waste and inefficiencies*. It is also important to evaluate customer demands to ensure that your organization is only meeting those demands and not creating more than what is necessary.

8. *Think systematically*: In a system, there are many different interconnected parts that work together (interfaces). It is important to understand the relationship between each of these parts because it will help you make better decisions. You should avoid taking on a narrow vision of your organization and get rid of any barriers that interrupt the flow of ideas and information.

9. *Create constancy of purpose*: Employees should be informed of the goals and mission statement of the organization from *day one*. This shouldn't just stop after day one, however. You should continue to emphasize these goals and principles every day going forward. Every employee should have an *unwavering certainty* of why the organization exists, where it's going, and how it will get there. Knowing this will help them align their own actions and goals with those of the company.

10. *Create value for the customer*: To create value for the customer you have to understand *what the customer* **needs** (and wants and expects). The value is

simply what that person is willing to pay for. Organizations must continue to work to understand the needs, wants, and expectations of their customers. An organization that stops delivering value to the customer is not sustainable over time. (Usage of the *Kano model* will facilitate the creation of value for the customer. The Kano model forces an organization to evaluate: (a) the basic, (b) the performance, and (c) exciting characteristics as well as tracking the movement from one category to the next. It is a dynamic process, so we must be careful.)

COMMON AND POWERFUL OPERATIONAL EXCELLENCE METHODOLOGIES

Through *operational excellence*, an organization can improve its *company culture and performance*, which leads to *long-term sustainable growth*. Businesses should consider looking past the traditional one-time event and move towards a more long-term system for change. Over the years, *many methodologies, standards, and guidelines* have been introduced to the mainstream business culture as a method of achieving operational excellence. Here we will look at the three most popular methodologies as identified by Kothari, A. at: https://tallyfy.com/sipoc-diagram/. Retrieved on April 25, 2020.

Lean Manufacturing

Lean manufacturing focuses on systematically minimizing waste in a production system. (For more information, see Anderson and Fagerhaug, 2000; Ward, 2014.) The reader should notice that we are talking about minimizing waste – NOT eliminating waste. To eliminate waste is an impossible task. So be careful how you can define your effort (operational definition, issue). It teaches that the only thing a business should focus on is that which *adds value*. Lean also teaches that every process has some sort of *bottleneck* and that focusing all your improvement efforts on that bottleneck is the quickest path to success. The key principles of lean manufacturing focus on *improving the quality of products and services*, minimizing anything that doesn't add value and reducing overall costs. Traditional lean manufacturing identifies seven areas of waste which are commonly referred to as the *seven deadly wastes*. Specifically, these are as follows:

a. *Overproduction*: Overproduction happens when employees produce something before it is actually needed. This is one of the worst forms of waste because it leads to excessive inventory and often masks underlying problems.
b. *Waiting*: When employees are left waiting for the next step in production, no value is being added. It can be very eye-opening to examine each step from the beginning to the end and then evaluate how much time is actually being spent adding value and how much time is spent waiting.
c. *Transport*: Transport is waste caused by unnecessary movement of uncompleted or finished products.
d. *Motion*: This step refers to all movement that doesn't add any value to the product and is usually caused by poor work standards.

e. *Over-processing:* This happens when more time is spent on processing than is necessary to produce what the customer needs. (Sometimes we call this the effect of diminishing returns.) It is also one of the hardest wastes to get rid of, because we think we are doing something.

f. *Inventory:* This type of waste occurs when the supply exceeds what the actual demands are.

g. *Defects:* Defects are mistakes that will either need to be fixed or the process will have to start over entirely. In manufacturing, this usually looks like a part that either has to be scrapped or completely remade.

SIX SIGMA

Six Sigma is a set of tools and techniques that are designed to improve business processes which will result in a better product or service. The goal of Six Sigma is to improve the customer experience by identifying and minimizing variation. Many of the Fortune 500 companies have implemented Six Sigma to some degree, claiming substantial savings. The savings come about from the fact that a Six Sigma business will produce no more than 3.4 defects for every million opportunities. A defect is defined as anything that fails to meet the customer's expectations. It does this primarily by implementing DMAIC. DMAIC is an acronym that stands for define, measure, analysis, improvement, and control. Here is a closer look at each step in this process and how it helps to build Six Sigma businesses:

a. *Define:* In this first step, the team will define the problem (the operational definition) because without knowing what the problem is, you really can't fix it. Once you have defined the problem you can begin creating a plan and evaluating your available resources.

b. *Measure:* Now that the team understands the problem, it needs to measure all of the available data and look closely at the current process. What is working well and what needs to be improved?

c. *Analysis:* Once the team has measured the data, the analysis begins to find the root of the problem.

d. *Improvement:* After the team analyzes the data, it begins coming up with possible solutions. Implement these solutions on a small scale to test the results so you can make any necessary changes. It is of paramount importance that the pending changes MUST be data driven and NOT politically correct resolutions.

e. *Control:* Once the team has implemented the new process, the team needs to find a way to maintain that process. Generally, this is done by utilizing some form of statistical process control (SPC). Continuous improvement is important to ensure that your process stays effective.

The DMAIC approach is used for assignable causes in the process of evaluating problems. For random variation issues, the define, characterize, optimize, and verify (design for Six Sigma – DFSS) is used. For a more detail description of the methodology, see Harry and Schroeder (2000), Pande et al. (2000), Breyfogle (2003), Eckes (2001), and Stamatis (2002–2003).

SIPOC MODEL

It is perhaps one of the most useful tools (diagram) in understanding the individual process. It is used (a) to kick-start problem-solving within the business process. It offers a linear visual presentation of Supplier – Inputs – Process – Output – Customer – see Figure 7.2. The intent is to use it so that operational definitions are understood and agreed by everyone. Perhaps, go as far as to standardize some key language of all the components of the model, (b) the SIPOC diagram may be the first step in creating a process map, and (c) the SIPOC aids the team in clarifying the process by asking relevant questions such as:

a. Who are the suppliers of the given process?
b. What requirements should the inputs fulfill?
c. Who are your true customers?
d. What specification do customers want for the end product?
   SIPOC diagram has a pretty straightforward structure. Its whole purpose is to present the INFORMATION at the core of the process in the simplest way possible. SIPOC diagram is one of the oldest and most trusted ways to map a business process in the most general way. It gives you a birds-eye overview of the process that could help you with onboarding a new team member or be the foundation of a future business process improvement initiative. To construct one, you can start with a table with five columns. Then, label each column with the letters SIPOC or the words Suppliers, Inputs, Process, Outputs, and Customers. A typical SIPOC model is shown in Figure 7.2.
e. *Start with the process:* If you decided to construct a SIPOC diagram, you probably already know which process you want to analyze. Write the name

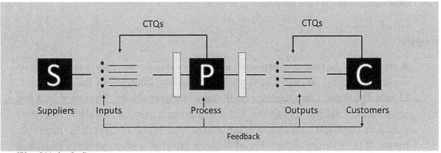

CTQs = Critical to Quality
P= Value added Process
C= Customer requirements final product
I= Inputs: Manpower, Machine, Material, Method, Measurement, Environment.

FIGURE 7.2   SIPOC model.

of the process (*this must be a value-added item*) into the middle column and briefly describe its key steps. You can either list them or draw a simple flowchart to make it easier to comprehend. When completing this step, keep a few things in mind:

    i. Make sure you know the exact starting and ending points of the process. If you don't, this can mess up the whole diagram once you move to the other columns.

    ii. Don't go into too much detail. Remember, SIPOC diagram is a high-level process map and is designed to get a birds-eye overview of the process. Do not include decision points or feedback loops.

f. *Identify the outputs of the process*: As with the previous step, focus on the key outputs of the process. In this step, write down the three or more main outputs. Use nouns for the most part and keep the tone neutral. Your goal is to avoid categorizing your outputs into good or bad ones – that's not the point of the diagram.

g. *Identify the customers*: In this step, list the people who benefit from the process. These don't have to be the literal "customers." For example, if you are working on a diagram for an internal process, the "customers" are your coworkers. Think of who benefits from this process. Who would be upset if the process is not complete?

    i. When doing the research for this step, up your game by noting customers' requirements in the "Output" column.

h. *List the inputs for the process*: Here you write down the inputs required for the process to function properly. Just like with every previous step, focus on the most important ones. The traditional 5M&E or some other specific inputs should do.

i. *Identify the suppliers of the inputs*: In the Suppliers column, write down the suppliers based on what inputs the process uses. Be sure to mention any specific suppliers whose input has a direct influence on the output. For example, imagine you're doing a SIPOC diagram for the process "Making tomato sauce." If the supplier has an impact on the variation of "Taste" output, you definitely want to list them.

KAIZEN

- Kaizen is a Japanese word and it means *continuous improvement*. It is used in business to implement positive, ongoing changes in the workplace. Generally, these changes are small steps but over a long time may become significant. Changes using the Kaizen approach may be 5%–10% improvements as opposed to reengineering which may result in over 30% improvements. The guiding principles of Kaizen are that a good process will lead to positive results, teamwork is crucial to success, and that any process can be improved. Organizations that have implemented *kaizen* have consistently demonstrated an improvement in creating a culture of continuous improvement. That means that (a) employees work together (team effort) to achieve ongoing workplace improvements; (b) using the process of *Kaizen* teaches that when applied consistently, small changes will compound over time and produce big results. These results may be in the areas of: employee productivity, cutting costs, and improving the customer experience; (c) One of the byproducts of the methodology is that it does not limit itself to small changes but it focuses on the participation of all employees to effect real change; and (d) *Kaizen* stresses the point that once improved, the cycle of improvement starts all over again for further improvement.

ACHIEVING OPERATIONAL EXCELLENCE (AOE)

Operational excellence (not to be confused with operational definition) is the ultimate goal of all organizations striving for continuous improvement. Projects and tools are a useful place to start but on their own, they are not enough to create lasting change. What is necessary for that excellence to occur management must be committed to provide the employees with the appropriate and applicable tools as well as demonstrate their own commitment to excellence in everything they do themselves.

So, the question is how do we get to AOE? We do it by constant improvement and using the Kaizen approach, Quality Function Deployment (QFD) to identify what customers want, benchmarking what do the "best in class" are doing and how we may incorporate some of their approaches, and process reengineering. All of these methods and many more can contribute to improvement and regenerate a stagnating company, by increasing the profits and driving growth. For more information on operational excellence, see Cartin (1999, 2010); for reengineering, see Manganelli and Klein (1994) and Stamatis (1997); and for QFD, see Day (1993).

- Reengineering (*a.k.a.* Business Process Reengineering – BPR). It is one of the many methodologies that creates a major change in the organization. As such, an auditor should be aware of some of the characteristics that make it worth the effort for a particular change. It is not an easy concept to comprehend, but the essence of the methodology is to tear down an existing process – something that everyone has become comfortable with – and then create something new. So, the steps for such an undertaken are:
  - *Step 1*: Identify and Communicate the Need for Change. The adage: if you do what you always did, you will always get, what you always got is appropriate here to be reminded of. Therefore, if something does not satisfy the organization and/or the customer, obviously, something must change. To make that change, it is imperative that that need for change must be identified and be communicated to all concerned.
  - *Step #2*: Put Together a Team of Experts. Any change depends on the contribution of several individuals. Reengineering is no different. Any effort for reengineering must have a team of experts who have knowledge about the pending (expected) change and be highly motivated. Typical initial members (later to be extended to others) must be cross-functional and multi-disciplined. The membership should consist of:
    - *Senior manager:* When it comes to making a major change, you need the supervision of someone who can call the shots. If a BPR

team doesn't have someone from the senior management, they'll have to get in touch with them for every minor change.

- *Operational manager:* As a given, you'll need someone who knows the ins-and-outs of the process – and that's where the operational manager comes in. They've worked with the process(es) and can contribute with their vast knowledge.
- *Reengineering experts:* Finally, you'll need the right engineers. Reengineering processes might need expertise from a number of different fields, anything from IT to manufacturing. While it usually varies case by case, the right change might be anything – hardware, software, workflows, *etc.*

This step is very important because if the team does not have the right composition, there will be a high probability of failure. That is why the cross-functionality and the multi-discipline compositions are essential for the success of the team.

- *Step #3*: Find the Inefficient Processes and Define Key Performance Indicators (KPI).

  Once you have the team ready and about to kick-off the initiative, you'll need to define the right KPIs. You don't want to adapt to a new process and THEN realize that you didn't keep some expenses in mind – the idea of reengineering is to optimize, not the other way around. While KPIs usually vary depending on what process you're optimizing, the following can be very typical:

- **Manufacturing**
  - *Cycle time:* The actual time spent from the beginning to the end of a process.
  - *Changeover time:* Time needed to switch the line from making one product to the next.
  - *Defect rate:* Percentage of products manufactured defective.
  - *Inventory turnover:* How long it takes for the manufacturing line to turn inventory into products.
  - *Planned* versus *emergency maintenance:* The ratio of the times planned maintenance and emergency maintenance happen.
- **IT**
  - *Mean time to repair:* Average time needed to repair the system/software/ app after an emergency.
  - *Support ticket closure rate:* Number of support tickets closed by the support team divided by the number opened.
  - *Application development:* The time needed to fully develop a new application from scratch.
  - *Cycle time:* The time needed to get the network backup after a security breach.

  Once you have agreed on the exact and the defined KPIs, the team will need to go after the individual processes. The easiest way to do this is to do business process mapping. While it can be hard to analyze processes as a concept, it's a lot easier if you have everything written down step by step. This is where the operational manager comes in handy – they make it marginally easier to define and analyze the processes. Usually, there are two ways to map out processes:

- *Process flowcharts:* The most basic way to work with processes is through flowcharts. Grab a pen and paper and write down the processes step by step. That write up should be conducted ONLY by visiting the process itself at the GEMBA and NOT from memory.
- *Business process management software:* If you're more tech-savvy, using software for process analysis can make everything much easier. There are many software available. Using such software might end up optimizing the said processes as it allows for easier collaboration between the employees.

- *Step #4*: Reengineer the Process(es) and Compare KPIs. Once you're done with all the analysis and planning, you can start implementing the solutions and changes on a *small* scale. At this point, what needs to be done is keep putting your theories into practice and seeing how the KPIs hold up. If the results show improvement, then scale up the actions with other processes. If not, you go back and start with new potential solutions.

A classic famous example of reengineering that has been written about in the literature is the case of Ford Motor Co. dealing with their accounts payable. Through a benchmarking study with American Express and Mazda, they were able to quantify their goal to reduce the number of clerks working in accounts payable by a *couple of hundred* employees. Then, they launched a BPR initiative to figure out why was the department so overstaffed. A pictorial view of the before and after process is shown in Figure 7.3.

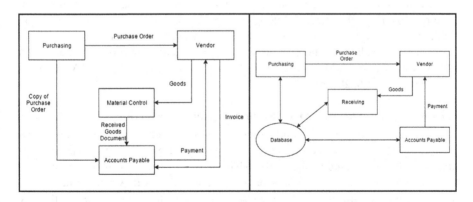

FIGURE 7.3 A pictorial view of the before and after reengineering process.

You can learn more about the case in Liang, Y. and E. Cohen. *Business Process Engineering: An Overview.* http://ccscjournal.willmitchell.info/Vol9-93/No5/Yong%20Liang.pdf. Retrieved on April 24, 2020. See also: Pearson, S. "Business Process Reengineering (BPR): Definition, Steps, Examples." https://tallyfy.com/business-process-reengineering/. Retrieved on April 24, 2020; Pearson, S. "Process Improvement (APQP)." https://tallyfy.com/apqp/. Retrieved on May 29, 20. In addition, see Cartin (1999, 2010), Hammer and Champy (1993).

STEPS IN OPTIMIZING *BUSINESS PROCESS REENGINEERING* (BPR)

Define and Plan the Program

Fulfilling customer expectations requires an understanding of what these expectations are. Thus, design and product reliability goals come first. Now, it's possible to formulate a preliminary bill of materials needed as well as a draft process flowchart that shows how the inputs will be transformed into outputs Next, it's important to define the product and process characteristics which in turn gives rise to a product assurance plan – an excellent methodology for this is the QFD. So, since management support is essential, the roles and responsibilities of management must also be determined. It will be an understatement to say that the suppliers must have a crucial role to play, and they can consider this portion of the process as being similar to Design and Development Planning and Design Input as specified in ISO 9001 and IATF 16949.

Designing and Developing the Product

To meet the goal of customer satisfaction, the company assembles a team to carry out a Design Failure Mode and Effects Analysis (DFMEA) to see what can go wrong and what the effect would be. The design and engineering specifications are also verified and reviewed in detail before a prototype plan can be finalized. If necessary, the design and its specifications may be adjusted. Meanwhile, the Advanced Product Quality Planning (APQP) team is looking at the equipment and tooling that will be needed. It will define key product and process characteristics, milestones needed while considering, and deciding on the testing requirements that will be used for quality verification and validation. Staffing requirements and management support are also on the agenda. It is imperative here to note that "the" APQP process can be used in simple and/or very complex processes in a way to satisfy the customers' requirements through a cross-functional and multi-discipline team. This interaction facilitates a generation of appropriate paths to improvement by using feedback from the different members to improve. For a detailed discussion on FMEA, see Stamatis (2003, 2019), AIAG (2008a), and AIAG/VDA (2019); for APQP, see AIAG (2008b) and Stamatis (1996, 2019b).

Aligning Manufacturing Processes for Customer Satisfaction

After reviewing the product and process quality system, the APQP team now focuses its attention on the elements that must be in place for productivity, efficiency, and uniform quality, all of which have a role to play in customer satisfaction. Items of concern here may be: a floor plan layout, process flowcharts, and a characteristic matrix – and since processes can fail just as designs can, they will carry out a Process Failure Mode Effects Analysis (PFMEA). Process instructions and a measurement system analysis plans are formulated, the capability of processes is studied, and packaging standards and specifications are drawn up. In addition, OEE, capacity, and contingency plans must be considered. For a detailed analysis of OEE, see Stamatis (2010).

Finally, It's Time for Product and Process Validation

Now that everything is working well on paper, it is time to see whether the process and product work as well as planned in the real world. This begins with a production trial run (part of the PPAP process). The production team implements the predetermined processes and measurement systems, and if all goes well in the final analysis, the product, its components, packaging, and all the accompanying control plans can receive final approval. For a very detailed discussion on PPAP, see AIAG (2009).

ASSESSMENT AND CORRECTIVE ACTION

Few plans, even the ones you've tested numerous times, will be without room for improvement. Once production begins in earnest, there are almost sure to be small issues that can be corrected or improved. Quality depends on the ability to produce predictably uniform products, and customer satisfaction, the goal of the whole of the process, will be an indicator that points towards any changes or improvements that must still be made.

If there's a marked need for corrective action, the entire process, or parts of it can be repeated, allowing for product changes that will meet client expectations. By looking through the steps involved in APQP and specifically design and process FMEAs, you probably have a good idea of why it's useful to use for your business. However, capturing them in a list helps us to fully grasp just why APQP and FMEAs are so helpful:

- The process keeps customer satisfaction at the core of its activities. Ultimately, resources are directed with the specific intention of fulfilling customer needs.
- Introducing a product becomes a carefully tested and validated process, limiting the number of changes that must be made (and fires to put out) after the product has been released onto the market.
- When faced with an urgent problem, the solution implemented may be the quickest but not necessarily the most cost-effective one. By pre-empting quality problems and firmly establishing client expectations, the process gives planners time to look for cost-effective solutions before the product is launched.
- By implementing APQP and FMEA, the company mitigates the risks inherent in introducing or modifying a product. (In most cases, ahead of time – which is by definition prevention.)
- Design and process specialists work together. This ensures that there is no miscommunication between them and that processes will fulfill the requirements of the design. If there is, it is enough time to either mitigate the difference or eliminate it completely.
- Suppliers can be briefed in accordance with APQP and FMEA guidelines and specifications. They, therefore, know exactly what is required, and how quality will be measured. (Remember, there are three approaches

to FMEA: (a) the AIAG, (b) the Ford Motor co., and (c) The VDA. So, make sure everyone is using the same guidelines for they are not all the same.)

- Manufacturing and assembly processes are carefully aligned with design specifications, limiting variation, promoting efficiency, and ensuring quality.
- Finally, APQP and FMEAs are process methodologies of continuous improvement that do not end once the company begins to produce and distribute a product. Feedback is received, and corrective action is taken.

Prevention and Diffusion

Quite often most organizations stop their journey to excellence with assessment and correction of "the" problem. It is unfortunate that this happens because unless you have a prevention plan to avoid the problem from happening again, it will appear again and again and you will end up fixing the same problem over and over again. Of course, there is no value in doing that. What is needed is a formal prevention plan that will not allow the problem to be repeated again. To do that, the escape point must also be identified as part of the corrective action. Once the corrective and prevention actions have been identified, there must be a strategy to identify the road blocks that will be perhaps prevent the solution to be implemented. In addition, a diffusion strategy of similar processes throughout the organization must be able to use that information because more likely than not, they will have either the same problem or a very similar problem to be content with. (An escape point is the point where the problem originated but was not caught. The investigation then has to identify why it was not caught and what can be done to "catch it," if it happens again.)

Statistical Process Control (SPC)

SPC is a method of quality control which employs statistical methods to monitor and control a process. This helps to ensure that the process operates efficiently, producing more specification-conforming products with less waste (rework or scrap). SPC can be applied to any process where the "conforming product" (product meeting specifications) output can be measured. Key tools used in SPC include run charts, control charts, a focus on continuous improvement, and design of experiments. An example of a process where SPC is applied is manufacturing lines.

SPC must be practiced in two phases: The first phase is the initial establishment of the process with its specific characteristics, and the second phase is the regular production use of the process. In the second phase, a decision of the period to be examined must be made, depending upon the change in 5M&E conditions (Manpower, Machine – this includes: wear rate of parts used in the manufacturing process, *i.e.* machine parts, jigs, and fixtures, Material, Method, Measurement, and Environment).

An advantage of SPC over other methods of quality control, such as "inspection," is that it emphasizes early detection and prevention of problems, rather than the correction of problems after they have occurred. In addition to reducing waste, SPC can lead to a reduction in the time required to produce the product. SPC makes it less likely that the finished product will need to be reworked or scrapped (AIAG, 2005).

Measurement System Analysis (MSA) A measurement systems analysis (MSA) is a thorough assessment of a measurement process and typically includes a specially designed experiment that seeks to identify the components of variation in that measurement process.

Just as processes that produce a product may vary, the process of obtaining measurements and data may also have variation and produce incorrect results. An MSA evaluates the test method, measuring instruments, and the entire process of obtaining measurements to ensure the integrity of data used for analysis (usually quality analysis) and to understand the implications of measurement error for decisions made about a product or process. MSA is an important element of the PPAP process, Six Sigma methodology, and other quality management systems.

MSA analyzes the collection of equipment, operations, procedures, software, and personnel that affects the assignment of a number to a measurement characteristic. A MSA considers the following:

- Selecting the correct measurement and approach
- Assessing the measuring device
- Assessing procedures and operators
- Assessing any measurement interactions
- Calculating the measurement uncertainty of individual measurement devices and/or measurement systems.

Common tools and techniques of MSA include calibration studies, fixed effect ANOVA, components of variance, attribute gage study, gage R&R, ANOVA gage R&R, and destructive testing analysis. The tool selected is usually determined by characteristics of the measurement system itself. In the automotive industry, the AIAG (2010) manual is followed for both the variable and attribute data. It is highly recommended that one should use the ANOVA approach as it identifies interactions of the measurement system. An introduction to MSA can be found in Montgomery (2013), Wheeler (2006), Niles (2002), Burdick et al. (2005), AIAG (2010), and Stamatis (2016).

Equipment: measuring instrument, calibration. This includes: wear rate of parts used in the manufacturing process, *i.e.* machine parts, jigs, and fixtures ration, fixturing.

- *People:* Operators, training, education, skill, care.
- *Process:* Test method, specification.
- *Samples:* Materials, items to be tested (sometimes called "parts"), sampling plan, sample preparation.
- *Environment:* Temperature, humidity, conditioning, pre-conditioning.
- *Management:* Training programs, metrology system, support of people, support of quality management system.

These can be plotted in a "fishbone" Ishikawa diagram to help identify potential sources of measurement variation. If ANOVA is used, the output is presented in a variety of graphical forms and charts.

Failure Mode and Effect Analysis (FMEA) FMEA often written with "failure modes" in plural is the process of reviewing as many components, assemblies, and subsystems as possible to identify potential failure modes in a system and their causes and effects. For each component, the failure modes and their resulting effects on the rest of the system are recorded in a specific FMEA worksheet. There are numerous variations of such worksheets. An FMEA can be a qualitative analysis but may be put on a quantitative basis when mathematical failure rate models are combined with a statistical failure mode ratio database. It was one of the first highly structured, systematic techniques for failure analysis. It was developed by reliability engineers in the late 1950s to study problems that might arise from malfunctions of military systems. An FMEA is often the first step of a system reliability study.

A few different types of FMEA analyses exist, such as

- Functional
- Design
- Process.

Sometimes FMEA is extended to FMECA (failure mode, effects, and criticality analysis) to indicate that criticality analysis is performed too. (Criticality is the product of severity and occurrence.)

FMEA is an inductive reasoning (forward logic) single point of failure analysis and is a core task in reliability engineering, safety engineering, and quality engineering. It is a bottom-top approach as opposed to fault tree analysis which is top-bottom.

A successful FMEA activity helps identify potential failure modes based on experience with similar products and processes – or based on common physics of failure logic. It is widely used in development and manufacturing industries in various phases of the product life cycle. *Effects analysis* refers to studying the consequences of those failures on different system levels.

Functional analyses are needed as an input to determine correct failure modes, at all system levels, both for functional FMEA or Piece-Part (hardware) FMEA. An FMEA is used to structure mitigation for risk reduction based on either failure (mode) effect severity reduction or based on lowering the probability of failure or both. The FMEA is in principle a full inductive (forward logic) analysis; however, the failure probability can only be estimated or reduced by understanding the *failure mechanism*. Hence, FMEA may include information on causes of failure (deductive analysis) to reduce the possibility of occurrence by eliminating identified *(root) causes*. For a very detail analysis and explanation of FMEA, see Stamatis (2003, 2019), AIAG/VDA (2019), and AIAG (2008b).

Advanced Product Quality Planning and Control Plan (APQP & CP) APQP and CP have become industrial standards (through the AIAG consortium) required by the OEMs to define the inputs and outputs of each stage of the product development process. APQP and Control Plans reduce the complexity of product quality planning for customers and suppliers by allowing customers to easily communicate their product quality planning requirements to their suppliers. Suppliers gain an understanding of basic industry requirements for achieving part approval from their customer. On the other hand, Control Plans summarize the identified process and product parameters required to maintain product conformity. These tools are applicable throughout the supply base in all customer/supplier relationships.

APQP serves as a guide in the development process and also a standard way to share results between suppliers and automotive companies. APQP specifies three phases: Development, Industrialization, and Product Launch. Through these phases 23 main topics are monitored. These 23 topics are completed before the production is started. They cover such aspects as: design robustness, design testing and specification compliance, production process design, quality inspection standards, process capability, production capacity, product packaging, product testing and operator training plan, among other items. In other words, one may characterize the APQP as a methodology that focuses on (a) up-front quality planning and (b) determining if customers are satisfied by evaluating the output and supporting continual improvement. It does this by evaluating progress through five phases, which are as follows:

- Plan and define program
- Product design and development verification
- Process design and development verification
- Product and process validation and production feedback
- Launch, assessment, and corrective action.

The APQP process has seven major elements:

- Understanding the needs of the customer
- Proactive feedback and corrective action
- Designing within the process capabilities
- Analyzing and mitigating failure modes
- Verification and validation
- Design reviews
- Control special/critical characteristics.

For more detailed information, see Stamatis (1998, 2019b; AIAG, 2008a).

Part Production Approval Process (PPAP) Production Part Approval Process (PPAP) is used in the automotive supply chain for establishing confidence in suppliers and their production processes. Actual measurements are taken from the parts produced and are used to complete the various test sheets of PPAP. "All customer engineering design record and specification requirements are properly understood by the supplier and that the process has the potential to produce product consistently meeting these requirements during an actual production run at the quoted production rate." *Version 4, 1 March 2006.* Although individual manufacturers have their own particular requirements, the Automotive Industry Action Group (AIAG) has developed a common PPAP standard as part of the APQP – and encourages the use of common terminology and standard forms to document project status.

The PPAP process is designed to demonstrate that a supplier has developed their design and production process to meet the client's requirements, minimizing the risk of failure by effective use of APQP. Requests for part approval must therefore be supported in official PPAP format and with documented results when needed.

The purpose of any PPAP is to

1. Ensure that a supplier can meet the capacity, manufacturability, and quality requirements of the parts supplied to the customer
2. Provide evidence that the customer engineering design record and specification requirements are clearly understood and fulfilled by the supplier
3. Demonstrate that the established manufacturing process has the potential to produce the part that consistently meets all requirements during the actual production run at the quoted production rate of the manufacturing process.

Critical elements of the PPAP

- Design record with all specification
- Authorized engineering change number (AECN)
- Customer engineering approval
- Process is clearly defined and understood
- Process is documented (verified and validated)
- Linkages of process are established (flowchart)
- Process is monitored, analyzed, and improved based on data (usage of SPC)
- Records are created, maintained, and retained
- Validation test report
- Control plan
- Capacity analysis
- PFD (current process flow diagram)
- Lab test report
- DFMEA (Design Failure Mode and Effective Analysis)
- PFMEA (Process Failure Mode and Effective Analysis)
- MSA study
- SPC.

Suppliers are required to obtain PPAP approval from the vehicle manufacturers whenever a new or modified component is introduced to production or the manufacturing process is changed. Obtaining approval requires the supplier to provide sample parts and documentary evidence showing that

1. The client's requirements have been understood;
2. The product supplied meets those requirements;
3. The process (including sub suppliers) is capable of producing conforming product;
4. The production control plan and quality management system will prevent non-conforming product reaching the client or compromising the safety and reliability of finished vehicles.

PPAP may be required for all components and materials incorporated in the finished product and may also be required if components are processed by external sub-contractors. The term ISIR (initial sample inspection report) is being used by German companies like VW and BMW. ISIR form is standardized by Verband der Automobilindustrie e. V., (short VDA), a German interest group of the German automobile industry, both automobile manufacturers and automobile component suppliers. The term is also used by some other companies like Hyundai and Kia. In fact, ISIR is like the Warrant and Inspection Report of PPAP document package.

PPAP document package includes some other documents such as PFMEA, control plan, drawing, MSA, and capability data. Besides ISIR document, other documents like that of PPAP are normally required by Volkswagen and Hyundai for release of a product and process. The PPAP is like the older ISIR plus much more, unless your customer has a specific requirement for using their ISIR within their system. ISIR is a summary of the initial sample being presented at whatever state.

The PSW is supported and validated by the ISIR. This does not mean the product being presented is under serial conditions but just states with evidence the current status. PPAP is the confirmation that the product meets the customer requirements for series production. The PPAP will be considered signed when a full PSW is approved by your customer and added to the PPAP folder. The PSW would always be supported with an ISIR but the PPAP is only considered approved when a FULL PSW is endorsed with an ISIR.

In essence, the PSW and ISIR are part of PPAP or VDA and can even be outside of PPAP in cases such as first off tool parts which should be submitted in most cases with a PSW and ISIR but will not be approved in PPAP until series conditions are met.

Sampling   Sampling is the selection of a subset (a statistical sample) of individuals from within a statistical population to estimate characteristics of the whole population. Statisticians attempt for the samples to represent the population in question. Two advantages of sampling are lower cost and faster data collection than measuring the entire population. Of course, ALL samples have a margin of error; however, that error can be determined ahead of time and controlled.

Each observation measures one or more properties (such as weight, location, and color) of observable bodies distinguished as independent objects or individuals. In survey sampling, weights can be applied to the data to adjust for the sample design, particularly in stratified sampling. Results from probability theory and statistical theory are employed to guide the practice. In business and medical research, sampling is widely used for gathering information about a population. Acceptance sampling is used to determine if a production lot of material meets the governing specifications.

There are many types of sampling and it behooves the auditor to at least be familiar with some of them and ask questions how the sampling is used in a given organization. Certainly, the auditor must be aware with the two kinds of errors: (a) the producer's error (Type I or alpha (α)) and (b) customer's error (Type II or beta (β)) that may determine acceptance of a "lot" or not.

Some sampling methods are mentioned here:

1. Sampling methods (there are several choices. Depending on the situation any one of these is acceptable: simple random sampling; systematic sampling; stratified sampling; probability-proportional-to-size sampling; cluster sampling; quota sampling; minimax sampling; accidental sampling; voluntary sampling; line-intercept sampling; panel sampling; snowball sampling; theoretical sampling.)
2. Replacement of selected units
3. Sample size determination (steps for using sample size tables)
4. Sampling and data collection
5. Applications of sampling
6. Errors in sample surveys (sampling errors and biases; non-sampling error)

PEST AND SWOT ANALYSIS

Everyone wants to maximize the effort in identifying the value of specific projects with the intent of having a measurable improvement. However, because maximization is very difficult – if ever possible, to accomplish a good approach is to focus on optimization which implies allocation and analysis of resources for any good decision. Two easy strategic methodologies for identifying improvement projects and define specific objectives are the PEST AND SWOT methodologies.

From an auditing perspective, it is appropriate to think about optimization by thinking the *SMART* model. SMART of course is an acronym which stands for **S**pecific, **M**easurable, **A**chievable, **R**elevant, and **T**imely. An evaluation and review of any initiative or specific project with focus on *continual improvement* will more likely miss the goal, unless some strategic discussion takes place. So, how does the SMART model work? The following overview summarizes each of the steps:

- *Specific*: Specificity of what is expected is fundamental to planning. Therefore, the definition of the goal must be clear and concise. To do that, one must define the:
  - *Who:* Who is the target and who is responsible?
  - *What:* What is the expected outcome? Are the outcomes in agreement with the organization's goals?
  - *Where:* Where is the implementation going to take place?
  - *When:* When the implementation will begin and when will it finish. Is the timing appropriate?
  - *Why:* Why is this objective selected? Is it really a priority for improvement or is it an internal political issue to display power?
- *Measurable:* Are the results measurable? Significant?
- *Achievable:* Are the goals "doable?" Are they realistic?
- *Relevant:* Are the goals "really" focused on improvement? Do they align with the overall strategic plans of the organization?
- *Timely:* Is the overall timing good enough to meet "key milestones" of the overall project?

These concerns may be delineated via a PEST analysis and/or a SWOT analysis. The PEST analysis is used very early on in the identification process for any political, economic, social, and technology that may be of future consequence, if the project is implemented. A PEST analysis for all intents and purposes is an analysis to bring together the four environmental perspectives to strategic planning. A typical format of the analysis is shown in Table 7.1.

On the other hand, the *SWOT* model is used when the issue of risk is at hand. It is also an acronym and it stands for: **S**trengths, **W**eaknesses, **O**pportunities, and **T**hreats. It uses a template for the analysis which fundamentally addresses the question of: how can the organization leverage its strength and weaknesses to take advantage of the opportunities while addressing or even mitigating risks. So, the SWOT analysis will allow the team to come up with some strategic goals to minimize risks to the project at hand. A typical format of the analysis is shown in Table 7.2.

TABLE 7.1
A Typical Format of PEST Analysis

Political	Economic
The political perspective should consider at least: political stability (especially for foreign investments or improvements in foreign organizational facilities), laws, taxes, and employee availability (qualifications)	The economic perspective considers market condition, inflation, interest rates, unemployment rate and if the organization deals in foreign lands, currency rates (exchange)
Social	**Technology**
The social perspective examines demographic and cultural factors	The technological perspective considers the current level of technology and future development

TABLE 7.2
Typical Format of SWOT Analysis

	Strengths	Weaknesses
Internal	Strengths are *all* the internal factors that the organization does well and will be available for the specific project	The weaknesses are *all* the internal factors that the organization does not do well or is not available to it (usually this is an issue of resources, *i.e.* people, machines, material, measurement, and method)
	Opportunities	**Threats**
External	Opportunities are *all* the eternal factors that an organization can leverage or use to achieve optimum results for the intended project	The threats are *all* the external factors that could potentially bottleneck or prevent the project to be completed on time and on budget

TURTLE DIAGRAM

The greatest risk to any process is the exchange or handoff between the processes. Most problems occur at the interface between activities or processes. This is why it is important to manage the system as a network of interrelated processes. Focusing on isolated functions misses the interaction(s) that occur between processes. The contribution of the knowledge to any interface action will more likely be closer to the root cause of the problem and thereby identify the "escape point." In understanding this interface, there is a tool that one may use, and it is called the *Turtle Diagram*. Indeed, the "Turtle Diagram" is a great tool for visualizing process characteristics. Of course, we all know that every process is made up of inputs, outputs, criteria, *etc.*, and so the Turtle Diagram provides the opportunity to visualize "a" process and to assist the effort to improvement. It takes its name because it looks like the body of a turtle with components as body (process); legs (who, what (resources, measures), and how); head (input) and tail (output). Specifically:

- Customer input requirements (tail)
- Customer output requirements (head)
- With what? What infrastructure and resources are needed to transform these inputs into outputs (Top Right Leg)?
- With whom? Any defined training or competency requirements? (Top Left Leg)
- How many? Output data used to measure process performance. (Bottom Right Leg)
- How? Any required process documentation? (Bottom Left Leg)

Even though there is no specific requirement of the ISO 9001, in clause 4.1, there is a strong implication of its usage with the stated verbiage of: *The organization shall identify the processes needed for the QMS and their application throughout the organization and determine the sequence and interaction of these processes.* So, it is to the organization's advantage to utilize the Turtle Diagram for both (a) communication tool of the process and (b) improvement of the process – especially the inputs and outputs. When the diagram is complete it facilitates the understanding of the process itself and as a consequence of this understanding, better measurement system can be developed for measuring the expected increase in efficiency. Furthermore, it is an excellent tool to evaluate the effectiveness for: (a) process design, (b) process FMEAs, and (c) process auditing.

With this cursory overview, let us examine with a short explanation its construction. The Turtle Diagram is made up of six areas, all surrounding the process, which is considered the turtle body. They are: (1) inputs, (2) materials and equipment (what), (3) support processes, procedures and methods (how), (4) outputs, (5) competence skills and training (whom), and (6) finally performance indicators (results).

- *Process:* The center of the diagram is titled "process." By definition a process MUST be a value-adding step.

- *Inputs:* This category should define the details of the actual process including: documents, materials, information, requirements, etc.
- *Outputs:* This should include details of the process such as products and documents.
- *Support processes, procedures, and methods:* Support materials include procedures, instructions, specific methods, etc. that contribute to the value of the process.
  - *Whom:* This section is dedicated to finding **all** the employees whose roles within the organization have the responsibility to value-adding steps within the process.
  - *What:* This section identifies *all* the resources needed to perform the process.
  - *How:* This section identifies *all* specific documents within the management system that tell the people responsible for completing the value-added steps how to successfully complete them within the organization's best practice.
- *Results:* This section looks at the measures the organization has at its disposal to monitor how well the procurement process is performing. If the measures align to the organization's plan, policies, goals, and objectives, then the measures should be able to tell management if the procurement process is fulfilling or needs to be adjusted or improved.

Auditing: The Turtle Diagram is also beneficial when it comes to auditing. The diagram allows the auditor to understand the input/output processes and activities under the scope of the internal audit. The diagram can help give an auditor a guide to perform a process audit and helps identify the direction of the audit it will follow.

An option of the full Turtle Diagram is a simplified version of the process model. In this approach, the four legs are combined into two legs known as Resources and Controls. An example is shown in Figure 7.4.

VISUAL FACTORY

Visual controls are a system of signs, information displays, layouts, material storage and handling tools, color coding, and poka-yoke or mistake proofing (MP) devices. These controls fulfill the old-fashioned adage: *a place for everything and everything in its place.* The visual control system makes product flow, operations standards, schedules, and problems instantly identifiable to even the casual observer (https://www.isixsigma.com/dictionary/visual-controls/. See also https://txm.com/visual-controls-workplace/. Retrieved on June 3, 2020).

Visual controls are an important part of management in every manufacturing and office workplace allowing a quick recognition of the information being communicated, in order to increase efficiency and clarity. When people come together to complete tasks and add value to the products or services, we need to find ways to clearly and efficiently communicate with each other. Here we will look at why we need *visual controls as part of our management strategy* and identify the two basic types of visual controls and how we initiate the implementation of these.

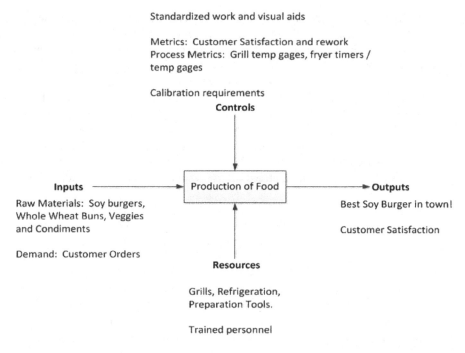

Standardized work and visual aids

Metrics: Customer Satisfaction and rework
Process Metrics: Grill temp gages, fryer timers /
temp gages

Calibration requirements
Controls

Inputs   Production of Food   **Outputs**
Raw Materials: Soy burgers,   Best Soy Burger in town!
Whole Wheat Buns, Veggies
and Condiments   Customer Satisfaction

Demand: Customer Orders
Resources

Grills, Refrigeration,
Preparation Tools.

Trained personnel

FIGURE 7.4   A simplified model of the Turtle Diagram.

The ultimate intent of all visual controls is to allow us to communicate without words and share information without interrupting. It helps to get everyone working together by providing a clear understanding of what is required at that point in time. Visual controls contribute to the management of every process in a way that individuals alone are not able to do, by showing where discrepancies occur. As with many of the lean enterprise principles we need to put systems in place to easily identify when things are going well so don't need to worry about them. This allows operators and the management team to easily assess the situation across the factory and identify the need to act when things are not under control or an appropriate response is not being undertaken.

The main principle that the visual factory operates under is: *it's hard to fix what you can't see*. Therefore, it is important that everyone in management, production, and support teams to be able to see (at a glance) what is going on and to know if it's good or bad. Implementing visual controls, reviewing them on a regular basis, and sustaining them will assist with the managing process, productivity, and certainly quality improvement.

To optimize this visual factory methodology, we use primarily two types of control. The first is display groups and the second is control groups. Controls or location controls include lines on the floor, color coding, and shadow boards for a range of processes. They are intended to guide the action of our team. This basic type of visual controls need very little explanation as to what they mean and what action is required. The challenge with these controls is ensuring compliance from everyone in the team especially the operators. The benefit of course with these types of control is that our brains are hardwired to comply, making the right thing to do being the only thing to do.

Display controls are charts that display metrics about the process with the intent of providing "fast" information and feedback on the performance of the critical elements. So, the intent of the visual factory is to make it easier for people moving across areas to understand the status and knowing how they need to respond to particular metric that is being monitored. Therefore, the best displays are: (a) the ones that are extremely easy to understand and (b) there is consistency across each chart and across each area within the business.

How do we go about generating and evaluating "the" specific visual factory metrics? There are at least three fundamental questions that need to be answered. (1) "What do I need to know to make the process better?" (2) "What do I need to share from other similar processes to further the improvement process? And (3) what questions am I frequently asked? This is a good sign that there is a legitimate problem and it needs to be fixed. For more information, see Ortiz and Park (2010).

5S METHODOLOGY

5S was developed in Japan and was identified as one of the techniques that enabled *Just in Time* manufacturing (Hirano, 1988). Two major frameworks for understanding and applying 5S to business environments have arisen; one proposed by Osada (1995) and the other by Hirano (1995). Hirano provided a structure to improve programs with a series of identifiable steps, each building on its predecessor. As noted by Bicheno (2004, Toyota's adoption of the Hirano approach was "4S," with Seiton and Seiso combined (https://en.wikipedia.org/wiki/5S_%28methodology%29. Retrieved on June 1, 2020).

The methodology is based on a very succinct outline of some basic words, which if they are followed, everyone will be better off and efficiency will be increased. The words are:

- *Sort (seiri* 整理 *-1S)*: This is the first word – or stage. It requires a red tag area containing items waiting for removal. So, Seiri is sorting through all items in a location and removing all unnecessary items from the location. The goals of this stage are:
  - Reduce time loss looking for an item by reducing the number of items.
  - Reduce the chance of distraction by unnecessary items.
  - Simplify inspection.
  - Increase the amount of available, useful space.
  - Increase safety by eliminating obstacles.

The implementation of this stage is very simple and it requires to

- Check all items in a location and evaluate whether or not their presence at the location is useful or necessary.

- Remove unnecessary items as soon as possible. Place those that cannot be removed immediately in a "red tag area' so that they are easy to remove later on.
- Keep the working floor clear of materials except for those that are in use to production.
- *Set in order (seiton* 整頓 *2S):* The second stage requires a simple floor marking (sometimes shown as *Straighten*). Seiton is putting all necessary items in the optimal place for fulfilling their function in the workplace. The goal of this stage is:
  - Make the workflow smooth and easy.

  The implementation of this stage is very simple and it requires to
  - Arrange work stations in such a way that all tooling/equipment is in close proximity, in an easy-to-reach spot, and in a logical order adapted to the work performed. Place components according to their uses, with the frequently used components being nearest to the workplace.
  - Arrange all necessary items so that they can be easily selected for use. Make it easy to find and pick up necessary items.
  - Assign fixed locations for items. Use clear labels, marks or hints so that items are easy to return to the correct location and so that it is easy to spot missing items.
- *Shine (seiso* 清掃 – *3S):* The third stage requires a cleanliness point with cleaning tools and resources. Seiso is sweeping or cleaning and inspecting the workplace, tools, and machinery on a regular basis. The goals of this stage are to
  - Improve the production process efficiency and safety, reduce waste, and prevent errors and defects.
  - Keep the workplace safe and easy to work in.
  - Keep the workplace clean and pleasing to work in.
  - When in place, anyone not familiar to the environment must be able to detect any problems within 50 feet in 5 sec.

  The implementation of this stage is very simple and it requires to
  - Clean the workplace and equipment on a daily basis, or at another appropriate (high-frequency) cleaning interval.
  - Inspect the workplace and equipment while cleaning.
- *Standardize (seiketsu* 清潔 – *4S):* The fourth stage requires a Seiketsu. It is a process to standardize the processes. It is used to sort, order, and clean the workplace. The goal is to:
  - Ensure the repetition of the first three "S" practices.

  The implementation of this stage is very simple and it requires to
  - Develop a work structure that will support the new practices and make it part of the daily routine.
  - Ensure everyone knows their responsibilities of performing the sorting, organizing, and cleaning.
  - Use photos and visual controls to help keep everything as it should be.
  - Review the status of 5S implementation regularly using audit checklists.

- *Sustain/self-discipline (shitsuke* しつけ *– 5S):* The fifth stage requires a shadow board (with tools' outline) and worker's movement that is being used in production floor. Shitsuke or sustain the developed processes by self-discipline of the workers also translates as "do without being told." The goal of this stage is to make sure:
  - That the 5S approach is followed.

  The implementation of this stage is very simple and it requires to
  - Organize training sessions.
  - Perform regular audits to ensure that all defined standards are being implemented and followed.
  - Implement improvements whenever possible. Worker inputs can be very valuable for identifying improvements.
  - When issues arise, identify their cause and implement the changes necessary to avoid recurrence.
- *Safety – 6S:* The sixth stage requires that the organization has a safety plan for both processes and employees. The goal is to provide a safe environment. That is prevent, minimize, or eliminate hazards.

  Safety is a late comer in the 5S methodology. It is the additional step which focuses on identifying hazards and setting preventive controls to keep workers safe during work operations. Many will argue that safety is incorporated into the other 5S items. That is correct. It is indeed part of the other 5S methodology, but it is very subtle. That is why it is profoundly important to identify it separately for it prioritizes the significance of the safe work environment and it implies – at least from a safety perspective – a stress-free and healthy atmosphere where all workers feel safe and secure. Of course, a clean and organized workplace can also make it easy to recognize and control potential hazards. However, by focusing on safety separately, it brings it to the forefront of importance without any hesitation or doubt.

  So, what does safety offer to the organization? Perhaps the most important is the fact of identifying existing hazards and those which are likely to be present in the workplace. Yes, all employees must be aware of the different types of workplace hazards and evaluate these hazards through risk assessments or a Job Safety Analysis (JSA). However, in many cases this is not happening and this is where having the extra step of safety is advantageous.

  To be sure, workers should also wear appropriate personal protective equipment (PPE: safety glasses, helmets, masks, insulated and anti-slippery gloves, etc.) as an additional protection to hazards which are difficult to control or cannot be eliminated. The use of PPE must be checked and other safety protocols must be disseminated to all workers through training and toolbox meetings.

  A toolbox meeting, or toolbox talk, is a short periodical consultation at work, intended to make everybody aware of the different safety aspects and dangers at the work sites to increase the safety in the workplace. Toolbox meetings are mandatory in many countries to obtain and maintain certain certificates. The secret to having successful toolbox talks is found in discussing the work situations that directly impact the personal safety and health of the employees. Never forget that a severe accident can cost an employee his health and his career, while it may cost the company millions in compensations and litigation (https://blog.archisnapper.com/what-is-a-toolbox-meeting/. Retrieved on June 1, 2020).

Achieving 100% safety is never an easy task, but through incorporating safety to the original 5S method – and *kaizen* – an organization can help improve working conditions. Workers can not only focus on completing tasks for the day but also habitually contribute to the overall workplace safety. To improve and sustain safety, it is suggested that the ISO 45001 be followed and preferably, get certified to it.

VARIETY OF APPLICATIONS

5S + Safety methodology is so versatile that it has expanded from manufacturing to a wide variety of industries including health care, education, and government. Visual management and 5S + Safety can be particularly beneficial in the automotive industry where chronic problems linger on with serious consequences.

MISTAKE PROOFING

There are many misconceptions about MP. First, the use of mistake proofing is used interchangeably with error proofing. Second, some organizations use error proofing to relate design issues and MP for process issues. In either case, as long as the definition of what you are trying to do is clear, it really does not matter what is called. The reason for this blanket statement is that the fundamental issue of this methodology is based on two essential attitudes about human behavior:

1. *Mistakes are inevitable:* Murphy's law: "Anything that can go wrong will go wrong."
2. *Errors can be eliminated:* Solutions can make it impossible for the error to occur. Defects introduce waste time and material. Therefore, they impact overall quality as well as customer satisfaction.

So, an error is any deviation from a specified manufacturing process or standard. Specifically, they: (a) are inaccurate or absent conditions necessary for successful processing, (b) are mistakes or improper actions, and (c) prevent any error leaving the process station. So, in the strictness sense it is possible to have an error in the process, but the process itself may correct it (*e.g.* in a multiple rolling mill, an error may be generated early in the process, but as the process continues to operate, the problem is fixed). If the error is not fixed internally, then it ends up to the next process as a *defect*. For a visual of this action, see Figure 7.5.

Error-proofing devices prevent and/or detect errors and defects. This means:

• At best, devices should prevent the ability to make an error.
• If the error cannot be prevented, the device should prevent the error from being turned into a defect.

FIGURE 7.5 The process of generating an error and ultimately the defect.

- If the defect cannot be prevented, the device should detect the defect before we add any more value to it.

On the other hand, a defect is the result of any deviation from product specifications that may lead to customer dissatisfaction. In other words, to be a defect:

- The product must have deviated from specifications
- The product does not meet customer expectations
- The product has left the process station.

In its simplest form, MP is a four-step approach to circumvent problems from happening. One, however, must understand that even with the MP application(s) there is no guarantee that the "fix" will be permanent, unless the "fix" deals with orientation issue(s). All other fixes may degrade over time or may not be sufficient for "the" 100% fix. After all, the MP may itself break or in some way not work as expected.

So, the four steps for an optimum MP are as follows:

1. *Awareness:* Having the understanding that a mistake can be made.
2. *Identification or forecast a prevention method for trouble spot(s):* Identifying the potential and planning the design of the product or process to detect or prevent it.
3. *Detection:* If severity and occurrence is possible, a method of detecting the error prior to the component leaving the station is utilized.
4. *Prevention (contain as many discrepancies as possible):* Not allowing the possibility for the mistake to occur in the first place. Design for manufacturability and/or process controls is the end game here. The goal of using one of these approaches is to NOT pass defect or errors to the next work station. Rather it is to provide traceability as an integral part of containment for repetitive issues, such as scrap, rework, and so many other issues that plague organizations.

ROLE OF THE AUDITOR IN THE MENTIONED METHODOLOGIES

To be very clear, we must mention that any auditor does not need to be an expert on any of the above methodologies. However, all auditors must have some rudimentary knowledge of all these – and maybe even some others – depending on the process and its complexity. This is very important because the auditor's function is to confirm and validate the integrity of the organization's system based on standards and specifications that a given organization is operating under. Furthermore, they have to make a judgment of the corrective action that was used to remove the problem. This subtle requirement to evaluate the validation as effective is perhaps one of the most essential tasks of the auditor, as ALL of these methodologies are mentioned either directly or indirectly in both the standards and the specific requirements.

8 Performance beyond Specifications

Since the late 1970s and early 1980s, quality has been the dominant factor in everything we do. Many standards, requirements, specifications, and guidelines have been developed to define, implement, control, monitor (efficiency in process), and measure (customer effectiveness) with the sole purpose to increase customer satisfaction. That includes: needs, wants, and expectations. To be sure, all these innovations in methodologies, standards, specifications, and so on have indeed increased the quality level – both in knowledge and in performance. However, presently there are many who are arguing that the quality levels expected have reached a *diminishing level of return*. In other words, the level of the current quality, even though it is not 100% of what customers need, want, and expect is good enough for the defined product or service at hand. To go beyond that, it is superfluous, unnecessary, and very costly. This revolutionary thought is based primarily on three arguments:

1. The increase of recalls in products that quality has been and continues to be emphasized. Especially in the automotive industry, the numbers are frightening since they are published by the National Highway Traffic Safety Administration (NHTSA). For the specific statistics, see: https://www.nhtsa.gov/sites/nhtsa.dot.gov/files/documents/18-3122_vehicle_safety_recall_completion_rates_report_to_congress-tag.pdf. Retrieved on June 13, 2020.
2. The drastic decrease for the Malcolm Baldridge National Quality Award. When the award was initiated in 1988 there were 66 applicants, from 1999 to 2009 there were on average 80 applicants, in 2010–2013 there were on average 25 per year. In 2018, there were no applicants at all. In Europe, the EFQMEA has the same trend. For an analysis of the demise, see Carvalho and Sampaio (2020, pp. 42–49).
3. The decline of the ISO certifications (including renewals) per documentation of the ISO itself. Part of the reason is that the standards are complicated, they are changed more often than clients have time to adjust to the previous editions and the freedom and power of auditors to audit items that are not part of the standards. For more information, see Desilva (2020).

The three reasons just presented are sheer excuses for not wanting to follow some kind of standards. Perhaps, the complexity is a factor, perhaps the cost, perhaps the time consideration is not allocated appropriately or whatever the reason is, we refuse to accept that quality is not a good endeavor to pursue. Whatever it is called or whether the supplier and/or customer calls it – any name, there will be constrains of some form, which will be called ALWAYS *specific requirements* that unless they are met, there will be no transaction. For that transaction to occur, there must be a win-win outcome. That means both parties will benefit financially. Period.

Why the discontent? We submit that the wind mill of frustration, the plethora of standards and requirements (sometimes, overwhelming) and continuing changes many do not understand what quality is and/or how it can help an organization. So, let us begin to demystify this issue. First, we all must understand that as "a" customer (both end-user and intermediate) they have many demands and they seem to increase ever-more-often, as such, so does our reliance on the supply chain. Failures in any part of the chain impact on our ability to satisfy these customer expectations. Therefore, supplier management becomes more strategic than it has been in the past.

Secondly, adding to this pressure is the fact that more and more of the value-adding activities of any organization are being outsourced to specialized firms (*i.e.* suppliers) who can do these activities more cost effectively. Beyond the outsourcing of manufacturing, these activities also include design and development, distribution, and critical services such as plating, heat-treating, and even assembly. Some of these suppliers are located far from the firm itself (*i.e.* in different States and even different countries), increasing the complexity of supply chain management. The end result is that there is a need to standardize those needs, wants, and expectations as much as possible and that is where quality enters into the picture.

To be sure, there are many ways to define quality. In fact, many became very famous pushing their own version as to what quality is (Juran, Deming, Feigenbaum, Crosby, Harry and so many others). However, one realistic and easy to follow is that quality is always defined by THE customer and not outside sources, no matter who they are. In that definition, the following apply: (a) performance, (b) delivery, (c) capability, (d) capacity, and (e) response to change and issues.

Now the questions become: How can anyone keep track of all these items? Can we assume that they will be delivered as specified? Can we inspect everything 100% upon delivery? The answer to these simple questions is definitely not. What we need is a validation process that is correct, accurate, and on time. That process is an audit. An audit that will evaluate the conformance to the customer's requirements whatever they are. To do that we also need an independent entity (registrar and an auditor) to do the validation. For a very detailed discussion and different points of view regarding audits, see VDA (2016); Brumm (1995); Clements, Sider, and Winters (1995); Cottman (1993); Keeney (1995a, 1995b); Peach (1994); Lamprecht (1992); Mills (1989a, 1989b, 1989c); Parsowith (1995); Russell (2000); Sayle (1988); Linville (1992); and MacLean (1993).

AUDIT ROLES

ISO 19011 makes it easy for us to define the roles of the audit organization, management, and execution. That means that the auditor must plan, conduct, and document the audit as directed by the audit team leader. The audit team has the overall responsibility for the audit. That means:

- Ensuring the efficient and effective conduct and completion of the audit
- Obtaining relevant background information

- Assisting in the selection of the audit team
- Preparing and communicating the audit plan
- Recognizing when audit objectives become unattainable and reporting the reasons to the client and the auditee
- Directing the activities of the audit team
- Serving as central spokesperson for the audit team
- Issuance of the draft/final audit report
- Making recommendations for improvements to the QMS.

The client (auditee), on the other hand, is the person who commissions and establishes the scope of the audit. Normally, this is the Audit Program Manager, but it doesn't have to be. The responsibilities of the client are as follows:

- Identifying the need for an audit
- Establishing the overall scope of the audit
- Making the initial contact and obtaining the auditee's cooperation
- Establishing the audit team
- Approving the audit plan, when part of the organization's policy
- Issuing the final audit report to the appropriate distribution
- Informing personnel within the audited organization/unit about the audit and its objectives, as needed
- Providing facilities as needed by the audit team
- Appointing guides, as needed, to the audit team
- Cooperating with the audit team and providing access to facilities, personnel, information, and records as requested by the auditors.

In addition to the specific requirements mentioned in the standards and/or the organization's specifications, the auditor must carefully verify, validate, and evaluate the existence and effectiveness of the following items:

- *Maintenance*
  - A planning process that achieves reliability excellence, with manufacturing, operations, and maintenance working together. Comprehensive tasks are developed and delivered that support Total Productive Maintenance (TPM).
  - The spare parts and their storage are managed.
  - The critical parts are identified.
  - The maintenance planning must cover machines and tools for preventive maintenance (PM) and where applicable the PM should exist and if not, it should be developed.
- *Training*
  - Training plan and timetable exists for each employee aligned to the business plan based upon job requirements and evaluations.
  - Training process is standardized and effective. Flexibility chart is updated for all operation.

- *Supply chain management*
  - Tier supplier targets are defined and their performance are tracked annual audits are performed and issues found (these may not have developed into problems yet) are tracked until closed.
  - Quality data is used in the sourcing decision process.
- *FMEAs*
  - Have – as applicable – a Design FMEA.
  - All operations must have been analyzed for risk using a Process FMEA. PFMEA training or workshops must be done by cross-functional teams, including manufacturing team member input. Risk should be prioritized by (a) severity, (b) criticality (severity × occurrence), and (c) risk priority number (rpn) – severity × occurrence × detection. If the RPN is used (not recommended) make sure the numerical values are consistently applied.
  - Failure modes are included/comprehended in the PFMEA (*i.e.* wrong parts, mixed parts, etc.).
  - Make sure that the failures are not effects or causes.
- *Capability*
- Capability reviews of process equipment with high risk or impact (based on the key characteristic designation system) are held to identify process capability (>1.33 P_{pk}) and appropriate and effective corrective actions. A reaction plan for non-capable process is present. Corrective actions are documented so that stability is monitored and used for process capability.
- *Capacity*
  - Capacity review and stream allocation analysis based on 5- and 6-day production rates with a margin of 10% minimum.
  - Review the OEE – aim for at least 85%.
- *Visual standards and controls*
  - Review the applicability of the visual factory.
  - Review the usage of the 5S.
  - Review for consistency ALL visual standards are common within the facility (*e.g.* multiple lines) as well as between facilities building the same platform or product line – this applies for global facilities, if applicable.
- *Visual/tactile/audible standards – communicated and understood*
  - Visual standards – if appropriate and applicable – are clearly communicated to the team members (or the individual operators) at the work station and incorporated or referenced in the procedures and/or work instructions.
  - Team and/or operators have been trained to identify the visual standards with appropriate training.
  - Visual standards must be easily able to differentiate *good from bad* parts and certainly they must satisfy customer requirements.
- *Error proofing verification*
  - All error proofing (detection) devices are checked for function (failure or simulated failure) at the beginning of the shift, if not, follow the process control plan (PCP).

- If appropriate and applicable the "Red Rabbits" parts are calibrated.
- Error proofing masters or challenge parts (if, and when) used must be clearly identified.
- Records of verification are available and reviewed as appropriate and applicable.
- Reaction plan is standardized and understood in case of error proofing device(s) malfunction.
- *Deviation management (bypass the standard and/or requirements)*
  - The organization plant must identify manufacturing processes and error proofing devices which can be placed in deviation or completely bypassed from normal operations.
  - ALL risks for all deviations and bypass items are evaluated and reviewed.
  - Standard work instructions are available for each deviation and/or bypass process.
  - Implemented bypass and deviations must be reviewed regularly as the goal is to reduce or eliminate any deviations and/or bypasses.
- *Measurement system analysis (Gage R&R, Gage Calibration)*
  - Gage capability (Gage R&R, linearity, bias, stability, etc.) of monitoring and measuring equipment is determined and the equipment is certified/calibrated with traceable standards at a scheduled frequency.
- *Development of process control (PFMEA, PCP, SW)*
  - PFMEA, PCPs, and standardized work (SW) documentation are comprehensive, sufficient, and flow one from the other.
  - Critical operations are identified with the appropriate symbol (*e.g.* ∇) at the operation and in the SW documentation (*e.g.* work instructions).
- *Process Control Plan Implemented (PCP)*
  - PCP checks are performed at the correct frequency and sample size.
  - Sample size and frequency must be determined based on risk and occurrence of the cause and ensures that the frequency adequately protects the customer by ensuring that the product represented by the inspection or test does not reach the customer before the results of these inspection or tests are known.
  - Sample and frequency must be reviewed on regular basis or as appropriate.
  - Checks are documented using the proper control method (*e.g.* SPC – control charts, check sheets.
  - Reaction plan(s) from the PCP are present, followed, and effective.
  - Effective PCPs must include prevention plan(s) in addition to corrective action(s).
  - Process-specific requirements are met and audit records kept.
  - Action plans with both corrective and prevention plans must be created to close the gap.
- *Process Change Control*
  - Plant processes are validated with a Gemba visit and the process flow diagram relative to changes in design and/or process (5M&E).

- The plant follows a documented change control process for customers and internal changes.
- The DFMEA and PFMEA must be updated as needed and reviewed at least once a year.
- All changes (engineering and manufacturing and when appropriate, suppliers may participate) must be the result of meetings with the appropriate departments.
- *Change control – production trial run (PTR)*
  - A reasonable sample size is used in production trial run (PPAP – AIAG; PPAP – Phase 0, Ford Motor Co.) based on risk and confidence level. (Usually 300 samples.)
  - Parts are contained, stored, and clearly identified prior to and after the PTR and PPAP approval.
- *Layered audits prevention plan*
  - Layered audits are in place to assess compliance to standardized processes, identify opportunities for continuous improvement, and provide coaching opportunities.
  - Layered audit owned by management audit plan must include multiple levels of management.
  - Audits are tracked and their results recorded.
  - All non-conformances must be followed up, corrected, and provide.
- *Process audit*
  - Audits are tracked and their results recorded.
  - All non-conformances must be followed up, corrected, and provide.
  - Using the SIPOC model evaluate each of the components and follow up on non-conformities – if they exist.
- *Standardized work*
  - All work is documented using a standard format and meets all safety, quality, all standard and organizational requirements.
  - Workplace organization is implemented (*e.g.* 5S: 5S is a workplace organization method that uses a list of five *Japanese* words: *seiri* (整理), *seiton* (整頓), *seisō* (清掃), *seiketsu* (清潔), and *shitsuke* (躾). These have been translated as "Sort," "Set In order," "Shine," "Standardize," and "Sustain." The list describes how to organize a work space for efficiency and effectiveness by identifying and storing the items used, maintaining the area and items, and sustaining the new order. The decision-making process usually comes from a dialogue about standardization, which builds understanding among employees of how they should do the work. In the United States during the last 10 years the list has been amended with an extra S for safety (safe).
  - SW has to be detailed enough to ensure that operation is performed on standardized way on each cycle.
- *Rework/repair/confirmation/tear down*
  - Repairs (on and off line) are compliant with approved SW.
  - Repaired, reworked, or replaced material is processed at a minimum through an independent repair confirmation (second person or machine after repair).

- Reintroduction of worked part includes all downstream checks in order to ensure that all control plan inspections and tests are performed.
- *Alarm and escalation*
  - Non-conforming product having sufficient alarm limits with escalation alarms are responded to according to the alarm and escalation processes (reaction plan).
- *Non-conforming material and material identification*
  - Team members have SW and understand what to do with non-conforming items.
  - Conforming items are identified, handled, and stored appropriately.
  - Non-conforming and suspect material must be identified, quarantine, evaluated for disposition.
  - An appropriate and applicable method of containment that must be established that is effective and will ensure a bad product will not reach the customer. (Third-party inspection is not the best way to containment.)
  - Traceability is applied according to acceptable and agreed-upon methods of the finished product and reworked parts when needed.
- *Team problem-solving process*
  - A well-developed standardized problem-solving process (e.g. 8D, 5-Why, Six Sigma: DMAIC, 3 × 5-Why) exists at all levels of the organization.
  - Formal problem-solving activities are initiated according to complexity of the issue or problem and specific criteria.
  - Issues are identified, root causes analyzed, and robust actions completed in a timely manner.
  - Problem-solving is driven at the team level and all teams are involved.
  - Leaders are actively involved coaching and guiding the process.
- *Andon system implementation*
  - A well-functioning Andon system is implemented in all production areas to support the team members when abnormal conditions occur and communicate relevant information. (An Andon system is one of the principal elements of the Jidoka quality control method pioneered by Toyota as part of the Toyota Production System and therefore now part of the lean production approach (Liker, 2004; Everett and Sohal, 1991). It gives the worker the ability, and moreover the empowerment, to stop production when a defect is found and immediately call for assistance. Common reasons for manual activation of the Andon are part shortage, defect created or found, tool malfunction, or the existence of a safety problem. Work is stopped until a solution has been found. The alerts may be logged to a database so that they can be studied as part of a continual improvement process. The system typically indicates where the alert was generated and may also provide a description of the trouble. Modern Andon systems can include text, graphics, or audio elements. Audio alerts may be done with coded tones, music with different tunes corresponding to the various alerts, or pre-recorded verbal messages.)
  - All operational areas of the organization are using the Andon process as intended and this shows tangible (measurable) results on the operating floor.

- *Inspection gates (verification, validation, and final inspection)*
  - Final inspection – of some sort – must be in place. It could be 100% if nothing else is available.
  - All items must be verified for checking and validated as appropriate.
  - All quality checks are included in SW.
  - Successive and more frequent checks may be required for complex or difficult problems.
- *Fast response process*
  - Exit criteria with appropriate timing and defined for closing issues.
  - There is an awareness of corrective actions for all applicable workers.
  - Fast response with predefined methodologies and/or tools (5D, Health Charts, tracking sheets, PFMEA, Control Plan, internal audit results, etc.).
  - As necessary and depending on the severity or magnitude of the problem meetings, reviews and the lie are mandatory and recorded.
- *Quality-focused checks*
  - High-risk items from critical operations have a quality-focused check performed each shift.
  - High-risk quality-focused items from customer feedback and problem-solving are included in internal audits or other pertinent sources.
- *Feedback/feed forward*
  - There is a feedback system. The system is appropriate and applicable to forward and backward information (verification and validation) from the final inspection station to manufacturing and vice versa.
  - Quality alerts are handled appropriately. Generally, they are issued by engineering for pending problems. They do have a time limit of approximately 90 days. If nothing happens during that window, then the alert reverts back to the original specification.
  - SREAs – Supplier Request for Engineering Approval must be initiated (generally) by suppliers (manufacturing). The turnaround is about 72 hours. A change is requested and engineering must approve it before it is implemented. If engineering does not approve it, the original design remains intact. It must be appropriately approved and recorded.
- *PFMEAs risk reduction and annual review*
  - Are the PFMEAs reviewed frequently to identify possible risk reduction?
  - Are the failures, effects, and root causes appropriately identified?
  - Are there more than one failure, effect, and root cause identified?
  - Are the FMEAs following a standardized format (AIAG, 2008b; AIAG/VDA, 2019) or customer-specific approach, *e.g.* Ford (2011)?
  - Action plans for top issues must be identified.
  - Recommendations, responsibility, and closure time must be included.
- *Material-handling process and FIFO first in, first out*
  - A plant FIFO or material-handling process is documented and practiced in all operations.
  - Visual aids assist in process flow. They are used if appropriate and applicable.

- *Shipping approval packaging events*
  - Material is shipped in the designated production container with proper labeling for regular production and all saleable build.
- *Safety*
  - Appropriate documentation is necessary to demonstrate that the organization is focused on safety.
  - Systems are in place to reduce safety risks.
  - Communication processes are in place to assure that safety concerns are discussed by all appropriate personnel.
  - A safety system exists to identify concerns that are being tracked, reviewed, and addressed on a timely basis.
- *Contamination requirements*
  - Contaminated product is secure and quarantined.
- Inspection gates (verification and validation stations including final inspection).

9 Quick View of Auditing

In the previous chapters, we discussed the requirements and the "process" of auditing. Quite a tedious task. We made a distinction between the second and third parties from the internal audits. That distinction is very important because fundamentally the second- and third-party audits focus on conformance and/or compliance to requirements. On the other hand, the internal audit is more comprehensive and therefore, demands more authority and responsibility of carrying the appropriate and applicable tasks but also an internal auditor needs more knowledge of the standards as well as the process being audited. The reason for this increase of responsibility is the fact that internal auditors are to identify issues, concerns, and problems, recommend and evaluate actions to possibly fix them, but more importantly, their role is to identify possible improvements in the QMS of the organization. In fact, this is the primary value of a good internal audit even though in most organizations is treated as "another item" checked in a long checklist of things to do.

So, because of any audit's significance this chapter will attempt to summarize the core of the requirements for any effective audit. The issues addressed here cover both internal and external audits. However, the reader will notice that the majority of the items mentioned here are strictly for an internal audit.

Perhaps one of the critical issues in auditing is the fact that an organization must be aware of the ISO requirements. It is the basic foundation for any auditing activity. It is the first step in auditing as it is required by all automotive OEMs for an organization to be appropriately certified. Many have claimed that implementing ISO 9001 for the benefits it will deliver in improved productivity, reduction in process waste, and management of risks have seen the bottom line improve with time (Guasch, Racine, Sanchez and Diop, 2007).

The next step in auditing is to be aware that other standards are involved such as the ISO 14001, ISO 45001, IATF 16949, and organizational-specific requirements. To be sure, these provide not only standardization but also many requirements for a company. For some, all these requirements may be overwhelming and certainly herculean tasks. To diffuse this complexity, this chapter will hopefully help. Our intent here is not to review the mentioned standards individually but to focus on the requirements from a very general perspective.

SUPPLIER QUALITY – BEYOND THE MINIMUM REQUIREMENTS

It has been said that Garbage In, Garbage Out (GIGO). In quality, it is profoundly important that ALL quality is based where it starts. That is: the supplier. So, an auditor must be cognizant of the supplier chain and look for the quality in not only specific requirement but also how does the organization view the chain. Therefore, until an organization moves beyond the thought process that supplier quality is only to satisfy the customers, the organization will never have breakthrough in quality improvement. Too often, supplier quality is only viewed as a necessary evil in our organization, because someone expects it or demands it. Breakthrough quality will only happen when management changes their attitude (mind set) and behavior as to how they view the role of supplier quality in relationship to improvement and productivity.

If the truth be told, all of us will agree that the focus in manufacturing has been and continues to be – regardless of the euphemisms used to: (a) reward those that solve problems not those that prevent them and (b) production rather than quality for its own sake. The emphasis is not on prevention rather it is on detection. It seems that quality is an afterthought and only if the customer complains.

Auditing is a process that more-less forces (voluntarily) an organization to make sure that standards, requirements and internal policies, procedures, and instructions are followed. It is an approach to make organizations to transform their quality paradigm from detection (corrective) to a more dynamic preventive methodology. It is this reason that a good auditor regardless of their own expertise must have at least a rudimentary knowledge of the process being audited but also the current standards:

- ISO 9001 – current standard
- IATF 16949 – current standard (which is an extension of the ISO 9001)
- ISO 19011 – current standard
- Automotive Core Methodologies, as defined by AIAG:
  - Production Part Approval Process (AIAG: PPAP, 2009)
  - Advanced Product Quality Planning (AIAG: APQP, 2008b)
  - Measurement System Analysis (AIAG: MSA, 2010)
  - Statistical Process Control (AIAG: SPC, 2005)
  - Failure Mode and Effects Analysis (AIAG: FMEA, 2008; AIAG/VDA, 2019)
  - Customer-Specific Requirements (CSR)
- ISO 14001 – current standard (if applicable)
- ISO 45001 – current standard (if applicable).

So, why is supplier quality important? Contrary to many managers and leaders of organizations, there is value in auditing and specifically following the requirements that customers demand from their suppliers. In order for this value to be identified and taken advantage of, quality must be implemented throughout the supply chain. Failures in any part of the chain will impact the ability to satisfy the customer's needs, wants, and expectations. Therefore, supplier management *must* become more strategic than it has been in the past. Therefore, some of the key critical requirements for a "good" supplier management system are as follows:

- Identify the process(es) that require appropriate action for improvement.
- Identify the standards that the organization must follow for continual improvement.
- Identify the focus and the targets that need to be addressed (relevant requirements, environmental and regulatory).

- Engage *all* individuals who are considered as stakeholders (encourage their input). Engage means free exchange of ideas and the use of teams that have authority and responsibility to propose and implement change. Engaging people within the process is essential to provide shared ownership in the results. It increases the probability of success.
- Engage in a process approach review focusing on interaction (interfacing) of processes.
- Commit the organization to a continuous improvement and pass that commitment to the supply basis.
- Data-based decisions are expected for the commitment to be sustained.
- Encourage relationship management practices. Integration of relationship in the management's requirement for "questionable" product or service must be an open and honest endeavor otherwise the *blame game* will not be fruitful. Avoid the word "I" and use the word "WE."
- Leadership must understand and be committed to continual improvement. If not, continual improvement is a euphemism and nothing else. It is the leadership that encourages *all* employees and empowers them to be committed to quality improvement. They must understand why change fails and why it is important to take the appropriate steps to correct and prevent any non-conformances. There are at least ten reasons why leaders fail to implement the appropriate change in their organization. They are as follows:
  1. Leaders underestimate the change challenge. Unfortunately, many leaders do not recognize that change takes time to think and execute with good results. When leaders assign additional projects to employees on top of their jobs that is a recipe for failure and not success. When that happens, it shows that leaders are *oblivious of the priority concept and indifferent about the employees' schedule* (both professional (his job) and personal (quality of life)).
  2. Leaders believe in an evolutionary change – change will happen in due time. Many leaders have the preconceived idea that just because they say something it is a good idea, it is a workable idea and in due time it will be incorporated into the organization's culture. Unfortunately, they seldom – if ever – take into consideration the scope and/or the difficulty that *a* change may present on the organization as a whole or a particular department. Sometimes a willful underestimate of resources will be assigned to an initiative with unrealistic deadlines. All these are signals for failure.
  3. Leaders are not trained or have an intuition of how to go about leading an effective change. They lack the appropriate and applicable skills. They are good in recognizing the need to change and what to do, but they have difficulty to apply that knowledge. It seems that they want to skip steps in the implementation process so that they finish early.
  4. Leaders force the change. **All** changes are intimidated and create fear. After all, by definition, when you propose a change, one admits, in relative terms, that something is wrong. A successful leader will always inform employees of the *pending change* and then proceed with a

win-win attitude to implement it. *It is imperative to communicate that the perceived notion of "something is wrong" is not always true.* Sometimes we need change to improve things, even though things seem to be working "OK" for the time being.

5. Leaders miss the focus of *why* the change is necessary and instead, prioritize their actions on *how* the change is to take place. It is imperative that those who push for *a* change must know *why* the change is important and it must be communicated to the stakeholders. If it is, they have to provide the appropriate and applicable resources – including the appropriate and applicable timetable. Many initiatives fail because either they are rushed or prolonged.

6. Leaders either ignore or they are unaware of the complexity that the change will result in. It is common for leaders to minimize the complexity of *a* change and as a consequence they do not see the difficulty of the change. The result is frustrated employees and misappropriated time, schedule (milestones), and appropriate resources for completion. When "a" change is introduced without conviction and enthusiasm from management, the general response of the employees is something like: here we go again…, well, this is a different flavor of the month approach…, we wonder how long is this going to last? And so on.

7. Leaders are blinded by their organizational authority and underestimate the resistance that the proposed change will bring into the organization. Change is frightful and the leader must be convincing for the need of change. Without convincing and enthusiastic arguments, employees may follow the *orders* for the change. However, it will be followed haphazardly, without conviction and it will become temporary.

8. Leaders are not aware of *how* to lead an *effective* change. An effective change is very difficult to be implemented. It takes a behavioral change but also a knowledge change about the process. All standards and specifications require *effectiveness*. However, for many leaders, the knowledge is gained through training, motivational speakers, and consultants, but the behavioral change still remains a difficult issue. Leaders are good at knowing what to do, but they fail the implementation of how to use that knowledge. Part of the reason is that they are arrogant and too proud in asking the direct employees that will be affected by the change as to what they really need as opposed to speculate their needs.

9. Leaders become stale (complacent) in their vision, because of past success(es). One may say they have become vaccinated against change. They are fearful of it. They avoid it. Quite often, leaders may know *what to do* and *how to do it* but they still do not accomplish the mission of the expected change. This is because they have completed a superficial assessment of the need and have avoided the underline reason(s) for the change, as well as, the implementation (corrective and prevention) actions.

10. Leaders fall into the same practices of the past even though they are not as successful as they could be. Not changing is a safe way to keeping

the *status quo*. However, if the leader knows or even suspects that something is wrong and there is no action taken to fix it, or the same remedy of the past or present is continued, we can use the Einstein expression that is *insanity*. It is the leader's function to break any form of resistance to enthusiastically promote change for improvement.

So, how can a leader influence the organization and satisfy the customer(s) while at the same time increase the profitability of his own organization? The answer lies in the definition of quality. To be sure, the definition of "quality" has evolved and defined differently over the years. However, every organization must adopt and modify any given definition (by gurus, standards, and organizations) to reflect the increase customer expectations and the organization's needs. However, the core ingredients of quality and the ones that auditors focus for everyone are as follows:

- Performance to and beyond specification requirements – **quality**
- Delivery performance including support of small lot manufacturing – **delivery**
- Capability to continually reduce costs throughout the supply chain – **cost**
- Ability to respond rapidly to new demands, products and issues – **responsiveness**.

Of course, the term quality by itself implies that something is conforming to requirements and meets performance requirements. Those parameters obviously can take many forms. In the domain of the word quality, we also can cover many other areas, such as variation reduction and risk. So, let us examine each of these separately.

PERFORMANCE TO AND BEYOND SPECIFICATION REQUIREMENTS – *QUALITY*

This means

- Ability to meet specifications. This characteristic is perhaps the best known.
- Ability to reduce variation around nominal for critical features. This is essential for special characteristics. If possible and cost-effective reduce variation even beyond the specifications. Why? Because ALL variations are waste. In order to do this, the ability to understand the level of variation in a process SPC and process capability studies are essential. Variation reduction, especially beyond the specifications, requires a disciplined and focused methodology. Suppliers that can (or are willing to) do this are in short supply and should be given serious consideration for future contracts.
- Ability to integrate their quality planning process into your quality planning process. With the increased use of outsourcing and the ever-increasing importance of cycle time reduction, the capability to integrate customer and supplier planning processes is becoming more critical.
- Ability to understand what's critical beyond the customer requirements.

- Ability to focus improvement efforts to reduce the variation in areas of importance, *e.g.* use Team Problem-Solving, SPC, Kaizen, and Six Sigma.
- Ability to use team-based APQP process.
- Ability to use product design and development (including FMEAs).
- Ability to have delivery of prototype parts.
- Ability to provide validation activities (their PPAP).
- Ability to support for your validation activities (your PPAP).
- Many of the current warranty issues experienced in the automotive industry stem from inadequate communication or understanding of system requirements between customers and suppliers. Remember that the APQP process is a team effort, and some of your critical suppliers belong to that team.

Performance is the driving force for the majority of organizations. Usually that performance is at the expense of quality, but it is not recognized for a variety of reason including the "quarterly earnings." It is unfortunate, but it is happening. Hopefully, internal audits will identify the fallacy *of performance* being the critical issue and perhaps they can transform the organization. After all, they may be able to produce but can they sell what they produced at the projected price or do they discount the price based on the rate of rejects and rework?

DELIVERY PERFORMANCE INCLUDING SUPPORT OF SMALL LOT MANUFACTURING – *DELIVERY*

Delivery performance (right part, at the right time, in the right quantities, properly labeled) is a critical process in the world of lean manufacturing. Capability in this area can sometimes be the major factor in determining which supplier to use, especially if you depend on a Just in Time system. After all, you may produce a great product at a real cost-effective competitive price, but if you cannot deliver it at the right time at the right place, it is worthless. So, the key ingredients for delivery are as follows:

- Delivery capability
- Support for small-lot manufacturing
- Just-in-time capability (including Kanban control)
- Quick changeover
- High equipment availability (total productive maintenance)
- Efficient and accurate inventory management
- Necessary information technology infrastructure and knowledge
- Location. (The location of the supplier affects more than just shipping costs. It also affects the development of long-term relationships. Face-to-face meetings and joint launch planning, problem-solving, and continual improvement are important and may not be possible if the supplier is located thousands of miles away.)
- Costs of shipments (including frequency considerations; premium freight; and mail)
- Complexity of shipments (complexity of packaging)
- Ability to interact face to face on issues

- Strategic sourcing issues
- Engineering, including launch planning
- Problem resolution.

CAPABILITY TO CONTINUALLY REDUCE COSTS THROUGHOUT THE SUPPLY CHAIN – *COST*

Cost reduction is not a nice thing to do anymore. In the face of global competition, it is a requirement. Many suppliers lack the capability to systematically reduce their costs. While they may be the lowest "price" producer, they may not be the lowest "total" cost producer. As a precaution to accelerating costs, an evaluation of these should be forecasted for the next 6 months, a year or 2 years from the present. The lowest price is not always the best choice. Always consider total life costing before you decide. The key ingredients to consider for cost are:

- Direct cost of the product (material, labor, standard overhead)
- Indirect or hidden costs (service calls, maintenance cost, *etc.*)
- Poor-quality costs (Often 20%–40%) of sales
- Rate of rework and rejects
- World class (Six Sigma companies) is <5% of sales.

ABILITY TO RESPOND RAPIDLY TO NEW DEMANDS, PRODUCTS, AND ISSUES – *RESPONSIVENESS*

The ability and willingness to partner with the customer is becoming one of the key differentiators between suppliers. To be competitive, the supply chain must embrace a "win-win" attitude towards the chain's suppliers and customers versus a "win-lose" attitude.

The *supply chain* includes all organizations associated with the flow of information, goods, and services from the raw material stage to the final delivery to the end user. *Supply chain management* refers to the integration and management of supply chain activities and organizations through partnerships and cooperative sharing of information and knowledge. On the other hand, ***Supply chain management seeks to optimize the profit for all of its members***. Given that in most industries the market price is defined by competitive pressure, this can only occur through joint cooperation, process optimization, and cost reduction. The key ingredients are as follows:

- Engineering process communication and partnering
- Advanced quality planning process
- Engineering change control
- Strategic sourcing issues
- Costs, logistics, materials requirements planning, forecasting
- Information sharing and management
- Problem resolution
- Complaints
- Improvements.

SUPPLY CHAIN PRINCIPLES

Optimization and long-term success cannot be realized unless key principles are understood and practiced by all members of the supply chain. Perhaps the most important one is that ALL money in the supply chain comes from the end user. This principle is supported by the following three:

1. Unless the customer is satisfied, no one profits. The only stable supply chain is one in which each member in the chain receives a reasonable profit.
2. Maximizing profits for only one member doesn't work. Supply chain management is about economic value added, not just cost.
3. Economic value added includes all services and features, not just the final product.

SUPPLY CHAIN STRATEGY

A successful supply chain strategy involves three steps. They are as follows:

Step 1: Optimize individual functions and processes.
Step 2: Optimize costs and processes across all members of the supply chain. Develop partnerships with suppliers. Develop a win-win program for all concern.
Step 3: Develop integrated goals and strategies across the entire supply chain. Trust and pursue long-term relationships.

If we look into these steps with a more "intense focus," we find that

- Downstream customers – including end users – are expecting better performance from their suppliers. The Kano model is helpful here (basic, performance, and excitement items)
- Quality is not defined as simply meeting specifications; it also includes
  - The ability to reduce variation in critical attributes
  - The ability to partner in the design of products and processes that optimize the customers products and processes
  - The ability to deliver on time, every time, using smaller and more frequent shipments of products to support JIT manufacturing
  - The ability to drive down poor-quality-costs through:
    - Proven root cause analysis and elimination
    - Proactive preventive actions
    - Responsive actions to customer complaints
    - Focused waste reduction capabilities
    - The ability to work with the customer to optimize processes and activities throughout the supply chain
    - Meeting these challenges requires
      - Careful evaluation and selection of suppliers who can provide these capabilities

- Active ongoing monitoring to ensure suppliers perform up to their capabilities
- Supplier development means, to address
  o Problems and issues
  o Process optimization
  o Improvements in supplier capabilities
  o The development of long-term relationships and trust.

CORE QMS REQUIREMENTS ARE CAPTURED IN

As an auditor you must become familiar with the sources of where the appropriate and applicable information may be found in the organization being audited. In addition to the source, as an auditor it is your responsibility to make sure that there is an integrated supplier QMS, so that, ALL requirements apply equally throughout the supply base. As an auditor you should be always observant for the existence of at least the following items:

- Process efficiency
- Quality planning, including FMEA, control plans
- Design control, including validation
- Focus on prevention
- Communications including internal and external
- Training and competence.

If any one of these does not exist, report it and make notes about it. They will be considered as non-conformances.

On the other hand, there are other requirements that are not necessarily focused on suppliers per se, but they are necessary for supply chain efficiency or quality. Certification to ISO 9001 and conformance to IATF 16949 will help push these requirements throughout the supply chain. Some of these requirements that any auditor should be aware and should be applied throughout the supply chain are as follows:

- Plant, facility, and equipment planning
- Contingency planning
- Change control
- Identification of special characteristics
- Procurement including product verification
- Supplier monitoring
- Production scheduling
- Identification and traceability
- Storage, inventory and delivery
- Statistical monitoring
- Customer satisfaction monitoring
- Internal audits
- Process and product monitoring

- Corrective and preventive action
- Continual improvement.

AVOIDING THE MOST COMMON PROBLEMS IN SUPPLIER SOURCING

As we all know, things go wrong when one least expects them. In other words, things sooner or later will not go right. Therefore, one should be vigilant at least on the following practices, so they may be avoided:

- Question the lead time. Is it appropriate? (Especially for sourcing suppliers. Yes, quite often that lead is 3–4 years in advance of job 1, but still it presents problems in capacity, delivery, and financial viability.)
- Question the specific direction of a supplier or any organization to investigate a possible area. This may indicate a bias on their part.
- Evaluate closely the "process" of screening suppliers. This should include a review of repeat "marginal" quality suppliers or suppliers who continually have production, delivery, and quality issues.
- Investigate and evaluate the "surprise" orders, as well as orders that fall below minimum bidding requirements, but they are frequent. (This includes purchases with credit cards and charged in expense reports.) This implies "hiding" something. This is usually called "back door" buying. It is imperative that all important bidding must be competitive – not only for price but for total cost, service, maintenance, and so on.

Once these items have been reviewed and evaluated, then the selection process for a "full" or "conditional" (for specific item) supplier must be initiated based on specific criteria. Some of that criteria are as follows:

- Standard criteria for standard products (shelf items)
- Unique requirements for critical buys and customized services
  - Design and development
  - Equipment purchases (including tooling requirements)
  - Major and critical production parts
  - Professional services.

We must not underestimate that any organization must and does have flexibility for their operations and practices. After all, there is no such thing as "one size fits all" criteria for supplier selection. Instead, the organization should have a set of generic criteria that can be adapted to suit the nature of the purchased product, material, or service. Some of the issues that should be considered for evaluation and or review are as follows:

- Dimensions of evaluation
  - Management/ownership
  - Financial strength and stability

- Production capability, including capacity
- Delivery capability, including support for small lot manufacturing
- Quality management systems and capabilities
- Engineering capability
- Prior experience in the industry
- Information technology infrastructure
- Management/ownership
  - Cultural fit – essential for long-term relationships
  - History of success of management team
  - Parent backing – access to resources where needed
- Higher risks with small family and minority-owned firms
  - Current customers
  - How many, and are they competitors
  - Dictates to some extent how loyal they will be to you
  - Confidentiality considerations
  - Financial strength and stability
  - Financial resources needed to complete contract
  - Current financial health
- Dunn & Bradstreet, Moody's, Standard & Poor's, *etc.*
  - Management accounting maturity (future viability both short and long term)
- Cost management capabilities (*e.g.* activity-based costing – ABC)
  - Part of IATF 16949 Risk Analysis requirements in clauses 6.1–6.3
- Production capability, including capacity
  - Age and quantity of manufacturing equipment
  - Level of automation – technology
  - Level of error and mistake proofing
  - Efficiency of plant layout (synchronous flow)
  - Process capability (ability to hold to tolerances)
  - Plant and equipment investments, expansion capabilities
  - Equipment/plant flexibility – dedicated vs. general purpose equipment
  - In today's just-in-time manufacturing environment, the capability of suppliers to support lean operations is often one of the most important criteria for selection. A close scrutiny needs to be evaluated on the allocation of production lines. Capacity analysis during the PPAP process is critical, and it must be completed before approval of a preliminary PSW.
- Delivery capability, including support for small lot manufacturing
  - Implementation of lean manufacturing concepts
  - Just in time pull systems
  - Kanban control
  - Visual controls and 5S
  - Quick changeover
  - Autonomous maintenance, TPM
  - Location, transportation modes available, sequencing capability

- Quality management systems and capabilities
  - Certification status (ISO 9001 and IATF 16949)
  - Use of SPC
  - Material identification and control methods
  - Containment and corrective action process
  - Poka-yokes and detection capabilities
  - Process monitoring and continual improvement
  - Internal audit program (this includes follow up of the findings and recommendations)
  - Supplier management and control
  - Use of standardized work and visual controls
  - Performance history and awards
- Engineering capability
  - Design capability and structure (if design responsible)
  - Quality of FMEAs (either the AIAG's, Ford's, or the AIAG/VDA's version)
  - Launch process control, including project management
  - Engineering competence
  - Development tools used (Autocad, GD&T)
  - Capabilities in DFM/DFA, Value Analysis
  - Change control, handling of ECRs, and drawings
- Prior experience in the industry
  - Automotive experience (APQP, MSA, PPAP, SPC)
  - Experience with automotive customers
  - Performance history (Q1, Targets of Excellence, and other preferred supplier awards)
  - Information technology infrastructure
  - Capability for electronic notifications and correspondence
  - Compatibility with your IT structure
  - IT system reliability and fault tolerance
  - Age of IT infrastructure, plans for the future
  - Price/cost
  - Overall cost structure
  - Need for overtime considering existing plant capacity
  - Cost management methods (ABC, COPQ)
  - Capacity for cost reduction
  - Cost of poor quality (COPQ) reduction programs
  - SPC, FMEA, Six Sigma, Value Stream Mapping capability, Value Analysis, DFM/DFA
  - Transportation and shipping costs (focus on premium freight/mail)
  - Fixed investment amortization method used for price.

SUPPLIER DEVELOPMENT AND IMPROVEMENT

The ability of your suppliers to correctly and promptly identify and eliminate root cause is a very important capability. Their responsiveness and willingness to apply horizontal countermeasures to other processes reflects on their attitudes towards problem resolution. What can an auditor do to help in this situation? Some thoughts for evaluating are as follows:

- Verify key process characteristics
- Verify effectiveness of containment actions (protect the customer)
- Verify the use and adequacy of root cause analysis
- Verify the effectiveness of the actions that were introduced to eliminate/ minimize root causes
- Verify the effectiveness of the tools used for identifying "the" cause of the problem
- Review and identify weak points of failure
- Verify the findings (from the above review) have been communicated and reviewed by management and appropriate action (responsiveness) has been taken.

COMPLAINT HANDLING

No one is free of problems. However, if any difficulty in any system is viewed as a problem, that can create a defensive approach and therefore a prolonged analysis. It is better to view any "difficulty" as an "opportunity" without focusing on the "who" but rather as to "why" and "how" did the difficulty happened. Some of the approaches that a good auditor should be looking for are as follows:

- 5 Why or 3×5 Why analysis
- Detection methods for root cause (focus on the escape point)
- Verification plan to validate corrective and preventive actions
- Horizontal application (solution(s) diffusion in the department, facility, and/ or organization)
- Documentation updates. Warning! It is great to have a problem solved; but if there is no updating to any relevant procedures/instructions, it will not be too long before the "old" method of doing things will creep into your process and the problem will appear again.

So, how do you make sure that these actions are taken? Perhaps as an auditor you have to use both observations and investigating methods that will verify the spoken actions. It is the auditor's function to ask questions and it is the auditee's responsibility to provide sufficient information to allow diagnosis. One of the great errors in auditing is the fact that most individuals are not forthcoming with answers and that is very unfortunate for the organization, since they will not find out the real problems. The auditee should be able to freely provide pictures, drawings, sketches, parts, and any pertinent information without hesitation and without fear of intimidation by their management. The idea of providing insight (background information) into impact of the problem will make solutions much easier to be identified and implemented. Obviously this forthcoming of information is based on the notion appropriate and applicable communication by all concerned. The intent here is to make the operators and all other participants "feel" as important and give them an opportunity to be an active participant as a stakeholder.

Third-party certification to ISO 9001, IATF 16949, other standard(s), or customer requirements does not alleviate the need to conduct periodic audits or assessments of your suppliers. Use their performance history and criticality of their products to guide you in your audit program frequency.

A TYPICAL QMS AUDIT

- On-site evaluations *after* source decision but before kickoff
  - Less sensitivity to statements regarding findings
  - More of a traditional evaluation
  - Share findings with the supplier but still put into a positive framework – can write non-conformances (NCs).
- Absolutely essential for critical parts or suppliers
- Set up a defined frequency
- Tie into Supplier Certification Program
- Consider using a Capability Maturity Model in addition to standard QMS audit (*i.e.* compliance-based)
- Use trained and experienced auditors who have been trained in auditing and the related core tools
- Do not recommend actions in haste, except for possibly immediate containment actions for product problems
- Use checklists with predefined topics for evaluation
- Prior to the audit review:
  - Supplier performance history
  - Details of the product provided by this supplier
  - Complaints, issues, and committed actions in response
  - Previous audit reports for this supplier
  - Audit process ISO 19011.

REQUIREMENTS PASS-DOWN

- Initial and ongoing pass down of key requirements
  - ISO 9001 and IATF 16949 conformance
  - ISO 14001 certification
  - Monitoring of manufacturing process performance
  - New customer-specific requirements (*e.g.* new PPAP)
  - Required improvements in the capability assessment scores.

PENALTIES AND DEBITS

- Typically includes direct material, direct labor, production lost time, transportation
- Consider including other hidden elements that make up the COPQ
  - Complaint documentation and reporting

- Complaint communication and tracking
- Verification activities
- Where possible calculate and assign a standard cost per occurrence
  - Standard administration and handling costs
- Build these into initial contracts, where possible.

More and more companies are calculating the hidden costs of defects and are invoicing these costs to their suppliers when a problem is the direct result of a supplier quality issue. Some are even developing standard complaint handling costs and putting these costs directly into contacts and terms.

INTEGRATED LAUNCH PLANNING

- Active participation in APQP activities
  - Definition of requirements
  - DFMEA/PFMEA joint development
  - Defining special characteristics
  - DFM/DFA and value analysis
  - Quality function deployment
  - Launch planning
  - Validation activities (including Capacity, PPAP)
- Design for Manufacturability and Design for Assembly (DFM/DFA)
  - Simplifying and optimizing product designs to make them more reliable and less costly
  - Requires in-depth understanding of the planned product, its features and functions, and manufacturing and assembly operations
  - Should be started at the product conceptual design stage
- Quality Function Deployment
  - Applicable to design and build suppliers
  - Can create superior products in half the time of traditional design and development projects
  - Translates voice of the customer into design specifications, part specifications, and manufacturing part parameters
- Defining special characteristics
  - Features that require special processing controls to minimize variation because of potential significant effect on safety, regulatory compliance, or functionality
  - Joint supplier and customer responsibility
  - Requires understanding of product, process, failure modes, and their effects.

FMEA TRAINING AND CONSULTING

- One of the most value-added activities in the APQP methodology/toolbox is the FMEA. However, it is also one of the most poorly conducted activity due to its function and time required for completion

- Summarizes all important failure modes in product or process, what can cause the failure, and what countermeasures (preventive and reactive) are in place or planned to minimize the customer risk
- Misunderstood by both customers and suppliers alike. As such, the training and/or consulting should make sure that they address: (a) the function, (b) failure, (c) effect, (d) root cause, (e) preventive/detection actions, and (f) recommendation(s)/action(s). It is imperative then that any review of an FMEA should focus on these because they give insight into how much effort is being applied to the development of failure mode and effects analysis. Work with any of the stakeholders, including the supplier, if your review identifies weaknesses in the level of detail or methodology used to develop the FMEA.

SUPPLIER CERTIFICATION

- Motivation to improve performance
- Benefits to supplier
  - Preference in future awards
  - Fewer on-site assessments or surveillance
  - Less documentation review and approval (*i.e.* PPAP)
  - Stronger partnerships in joint activities.

JOINT PROBLEM-SOLVING

- Many problems are the result of interactions between a supplied product and the customer's production systems
- Historical practice of putting all of the burden for resolving such problems does not work
- Optimal solution is the one that minimizes the overall cost to the value chain
  - Win-win versus win-lose
- Such problems require joint supplier/customer problem-solving
- Necessary conditions for successful joint problem-solving
  - Open and honest assessment of the nature of the problem
  - Trust between supplier and customer
  - Open and honest sharing of the cost of the problem
  - Open and honest sharing of the cost of the solution
  - Realization that overall costs to both supplier and customer will be minimized when the overall cost to the value chain is minimized.

VALUE CHAIN MAPPING

- Joint mapping of the interdependent processes shared by both customer and supplier
  - Order generation and release process
  - Inventory management and control
  - Distribution and shipping process

- Design and development process
- Engineering change control process
- Purpose is to improve understanding of how each organization works.

VALUE ANALYSIS (VA)

Value analysis is a systematic method to improve the "value" of goods or products and services by using an examination of function. Value, as defined, is the ratio of function to cost. Value can therefore be manipulated by either improving the function or reducing the cost. Both are important and they have to be understood and applied appropriately by

- Joint examination to identify areas of waste and cost
  - Can be used for both processes and products
- Also called value engineering when applied to product designs
  - Asks of series of questions that are then used to redesign products or processes
  - Value analysis is very similar to DFM/DFA. The major difference is that DFM is applied during the initial design process, while VA is applied primarily to existing designs.
- General questions (Process)
  - Does the activity add value (does it serve a necessary purpose in the absence of problems)?
  - Can it be eliminated?
  - Can it be combined with another activity?
  - Can it be simplified?
  - Are the reviews and approvals necessary?
  - Are the delays necessary?
  - So on...
- Phases of VA
  - Preparation phase
  - Information phase
  - Evaluation phase
  - Creative phase
  - Selection phase
  - Implementation phase.

SUPPLIER CERTIFICATION PROGRAMS

Note that these classifications are not mutually exclusive. In fact, achieving certified supplier status is normally a prerequisite (but not the only requirement) for achieving preferred supplier status.

Beyond the obvious quality performance that is achieved by certified suppliers, preferred suppliers should also have strong delivery performance, demonstrated capabilities for continuous improvement, solid financial standing, price stability, and a willingness to partner with the customer in areas of mutual interest.

Note that some experts reverse the hierarchy of the classifications above, using certified supplier as the top-most status. I choose not to, since certification is associated primarily with quality, while preferred status captures other, just as important considerations.

Supplier that whose demonstrated performance and cooperation makes it a preferred source for additional business and joint development projects.

APPROVED SUPPLIER

- A supplier who has been evaluated and approved for use by the customer through some certification standard and their requirements
- Successfully completed the evaluation process
- Product continues to receive some level on incoming inspection and testing
- Normally go onto an approved supplier list or database
  - Informs buyer/requisitioner that this supplier has been evaluated and approved for use – no need to repeat supplier evaluation and selection process.

CERTIFIED SUPPLIER

- Supplier that, through previous experience and qualification, can provide material of such quality that it needs little if any inspection or testing before release into production. This certification is always given by a third party.

CRITERIA FOR QUALIFIED SUPPLIER

- Example Criteria
  - Minimum score based on customer's requirements (common minimum score is usually 80 or 90) on the Supplier Assessment Survey (known as MSA)
  - No rejected shipments for quality rejections, previous 12 months
- May also use a consecutive lot basis
  - Average delivery performance of 95%, previous 12 months
  - The benefits to the supplier of qualified supplier status are preference in future contract awards and progression towards a preferred supplier status. You may also allow them to self-survey annual supplier assessments (retaining the right to follow-up yourself).

CRITERIA FOR PREFERRED SUPPLIER

- Example criteria
  - Minimum score (as per customer requirement) on the supplier assessment survey
  - No rejected shipments for quality rejections, previous 12 months
  - Average delivery performance of 98%, previous 12 months

- Demonstrated willingness to partner on cost reductions, continual improvement initiatives, quality planning, and/or joint strategic planning.

The benefits to the supplier of the preferred supplier status are prestige, a significant advantage in future bids, and the opportunity to learn from the customer during joint initiatives (*i.e.* more development investment by the customer).

Don't try to certify everyone. The costs and efforts (annual assessments, QMS audits, focused development) involved are not warranted for many of your suppliers. However, maintain the standards high, but do not make it too easy to achieve preferred status because that will dilute the prestige that "prefer" brings.

CANDIDATES FOR SUPPLIER CERTIFICATION AND PREFERRED STATUS

- High-volume suppliers
- Suppliers of critical parts and materials
- Strategic suppliers.

MAKE ALLOWANCES FOR NEW MATERIALS AND PARTS

- Provides a means for qualification of new materials or components without risk of losing certification or preferred status
- For these "trial" or "special" status parts, require incoming inspection or testing
- Once qualified, then consider the normal criteria to apply.

New, cutting-edge parts or materials will have some initial quality or delivery issues. Don't discourage innovation by penalizing a certified or preferred supplier for the bugs that will inherently exist in these advancements. Agree beforehand on the need for an exemption for new products and on the criteria that will be applied. Then establish the appropriate safeguards to protect yourself and your customers. An appropriate way to do this is to follow the supplier request for engineering approval (SREA) methodology.

CERTIFICATION MAINTENANCE

- Accomplished through ongoing monitoring and reporting of results
  - Note that suppliers must have timely or real-time access to performance measures in order to anticipate and react to issues that could lead to decertification
  - Annual supplier assessments
- Self-surveys
- Conducted by your audit team.

If you allow self-surveys (the supplier completes the annual supplier assessment), consider use of the Self-Scoring Adjustment method that most customers provide for your use. Ensure the supplier knows that you retain the right to conduct your own on-site audit to verify their scores. Periodically (*e.g.* typically once a year – for critical suppliers, however, in some cases once every 3 years) conduct an on-site assessment after receipt of the supplier's assessment. This will improve the accuracy of the supplier's results.

DECERTIFICATION PROCESS

Decertification here means either loss of certified supplier status or removal of preferred supplier status. Decertification should not be taken lightly, and it should be recognized as a failure on the part of both the supplier and the customer. Do not use decertification as a punishment but rather as a motivator to make improvements where they are required. Pay attention to the following:

- Phased decertification to motivate correction
  - Probationary period
  - Decertification
- Include steps required to recertify
  - Original certification criteria
  - Focused recertification
- Work with the customer/supplier to avoid decertification
- Consider classifying problems for purpose of decertification
  - Major quality issue
- Production Stoppage (quality or delivery issue)
- End-customer rejection (make sure you understand the requirements)
- Customer complaint issued due to supplier problem
- Minor quality issue
- All other violation to requirements (minor labeling, partial shipment, few minor quality issues)
- Results in a quality alert or quality information report
- Example criteria – probation
  - Any major quality issue(s)
  - More than six minor quality issues over previous 6 months
  - Deterioration in supplier assessment score (different organizations have different criteria)
  - Overall delivery performance falls below certain criteria (different organizations have different criteria)
- Example criteria – decertification
  - Recurrence of a major quality issue during probationary period
  - Failure to respond to a formal request for action
  - Being placed on probation a second time during any 12-month period
  - Inability to meet the requirements for recertification within 6-month period
- Example criteria
  - Satisfactory response to problem resulting in probation (this includes prevention tasks that will not allow the problem to reappear)

- Zero recurrence of original complaint or quality issue
- Achievement of required supplier assessment rating or delivery performance
  - Sustaining the required level of performance for a period of 3 consecutive months
- No quality issues
- Acceptable (this depends on the customer/supplier contractual agreement) delivery performance.

The reader will notice that the issues raised here are more convoluted than a mere audit requirement. This is so because as we already have mentioned the internal audit is a function of identifying problems within the organization, but more importantly to identify potential IMPROVEMENTS for the organization. To do that, the internal auditor has much more responsibility than a second- or third-party auditor (they are ONLY concerned with conformance and/or compliance). This is the reason why we have presented an outline of things that should be considered and investigated.

10 Process Approach to Auditing

PROCESS AUDIT

Over the years there have been many definitions of "process." However, in the world of quality, the word is used in the context of three definitions. All of them are quite formal and they are based on the following sources:

1. A "process" is a progressive forward movement from one point to another on the way to completion: the action of passing through continuing development from a beginning to a contemplated end: the action of continuously going along through each of a succession of acts, events or developmental stages: the action of being progressively advanced or progressively done: continued onward movement (Webster's Third New International Dictionary, unabridged. Merriam-Webster, 2002, http://unabridged.merriam-webster.com, April 9, 2006).
2. A "process" is a set of interrelated work activities characterized by a set of specific inputs and value-added tasks that make up a procedure for a set of specific outputs (ASQ Glossary, www.asq.org/glossary).
3. A "process" is a set of interrelated or interacting activities that transforms inputs into outputs (ISO 9000, Quality Management Systems – Fundamentals and Vocabulary, third edition, 2005, clause 3.4).

WHAT IS A PROCESS AUDIT?

A few years ago, when I was presenting information on process audits, a representative of a major corporation told me it had been doing *process audits* for several years. When I asked him to explain the auditors' techniques, he said they took samples of the product at the end of the line and then measured the length, diameter, and weight. Based on results, they approved or rejected the lot. Many individuals believe that this practice is indeed a "process audit." But is it really?

Is that what a process audit is to you? I hope not, because what was described is more like *final inspection* or a *product or service audit.*

In a different situation, 2 months ago, another representative of a major corporation explained to her auditors that in a process audit there was no need for the use of a checklist. She proceeded to explain thusly: "I sent a document used to audit one of my corporation's processes. The auditing documentation was not called a checklist but was a list of regulatory and industry requirements by clause for a specific process, such as labeling." It is interesting to note that her remark, no matter how one evaluates it, is a checklist.

Is auditing a process by clause requirements or elements a process audit? Probably not because most standards and regulations do not take into account the dynamic nature of a process. When the audit criteria are the specific requirements of a standard, only key elements or critical controls are verified. Requirements may include the following:

- Written procedures.
- Record keeping.
- Verification of customer orders.

A list like this does not include "the" dynamic evaluation of inputs being changed into outputs. That is why quality-related definitions of a process are usually those found in ISO 9000 or the ASQ Glossary, as well as general dictionary definitions, for they all support the dynamic nature of a process and what it would mean to audit a process (see "Definitions of Process" above; for a very exhaustive analysis of Process Audit, see VDA: Process Audit – Part 3, 2016).

AUDITING BY ELEMENT

Auditors use various auditing techniques to collect evidence based on the audit scope and objectives. Auditing a process or system by element is one way and it verifies compliance or conformance to requirements. The value in this type of auditing technique is the direct linkage to license, contract, or regulatory requirements. Auditing a process by element ensures people are aware of the requirements and the organization is adhering to them. It helps prepare employees for external audits using the same criteria.

Furthermore, auditing by element also ensures a *state of readiness* for compliance and/or conformance to external requirements. It is a management tool for sustaining conformance to safety, health, or environmental and quality requirements. Of course, this is good, but be aware that for management, this technique defines auditing in the cost of doing business category, which is very limited approach.

AUDITING BY PROCESS

Auditing a process or system using process techniques verifies conformance to the required sequential steps from input to output. Process auditors use models and tools such as simple flowcharts, process maps, or process flow diagrams. The common element of these charts is that they all have identified the individual steps of the process, as well as, the correct (current) flow of the process (Russel, 2005, p. 17, 2006; Arter, 2003).

In a typical process diagram, the boxes in the center could represent a flowchart of the sequential steps. Flowcharts typically identify inputs, people, activities or steps, measures, and outputs. The auditor normally gets this information from a procedure or flowcharts provided by the audited organization.

During the first part of the audit, auditors should record current customer names, order numbers, routing numbers, and project numbers so they can link and verify process steps during the audit. Remember that all processes have all or some of the following traditional elements – known as 5M&E:

- People involved.
- Equipment (machinery) needed.
- Environmental requirements.
- Measures to test or monitor.
- Methods to follow.
- Materials used or consumed.

The use of process techniques is a natural steppingstone from conformance to performance auditing. When collecting evidence, auditors also will observe performance issues that would be of value to management. This observation is very important because as we all know, what is written and what is being performed may not be the same. Observation diffuses this issue of concern. (Warning: This is a problem with the distance auditing, since the auditor is depended on the "angle" or *depth of vision* that video or picture may be taken and the decision will be made. We all know that those issues in photography can and quite often dilute the significance of a finding.) In any case, auditors should report process performance accurate and without bias. They should be interested in indicators that support improvement efforts. These indicators include the following:

- Waiting.
- Redoing.
- Deviating.
- Rejecting.
- Traveling excessive distances.
- Premium travel, mail, and delivery.

All process performance issues should support improvement programs such as lean, ISO 9001, Six Sigma, or any other program that the organization has in place for improvement.

LAYERED PROCESS AUDITS (LPAs)

The simplest definition of an LPA is: a methodology of auditing that reduces variation and defects through repeat verification that operators are following standardized processes for the purpose of continual improvement and reducing quality costs. These audits are short and fast, with a set of yes or no questions that take about 10–15 minutes to complete. Their simplicity makes them accessible to people unfamiliar with process details, including those from other departments. A cross-functional and multi-discipline LPA team identifies and updates audit questions based on items such as:

- Past quality issues or defects and resulting corrective action
- Process failure modes and effects analysis (PFMEA) forms
- Procedures critical to product quality and/or customer satisfaction.

Auditors participate at a frequency corresponding to their level of authority, with audits taking place multiple times as in: daily, weekly, monthly, quarterly, and sometimes yearly. The frequency depends on the organization, number of process deficiencies, and availability of resources.

Automotive original equipment manufacturers (OEMs) like GM and Fiat Chrysler require their suppliers to conduct LPAs, which are also recommended by Ford. In the automotive industry, generally, they are used as a proactive approach to reducing risk that makes management accountable for quality, LPAs also help demonstrate compliance with standards such as ISO 9001 and IATF 16949.

So, layered process audits (LPAs) are a quality technique that focuses on observing and validating *how* products are made, rather than inspecting finished products. LPAs are not confined to the Quality Department but involve all employees in the auditing process. Supervisors conduct frequent process audits in their own area, while higher-level managers conduct the same audits less frequently and over a broader range of areas. These audits also typically include integrated corrective and preventative actions taken either during, or, immediately after the audit (Best Practices in Creating a Layered Process Audit Program. https://www.ease.io/resources/what-are-layered-process-audits/. Retrieved on July 14, 2020).

LPAs help manufacturers and service providers take control of processes, reduce mistakes, identify the true cost of quality (COQ) – the hidden truth – which is one of the metrics that most organizations do not record and/or report accurately and improve both work quality and the bottom line. Organizations with robust LPA programs in place see significantly lower rework and scrap, fewer warranty holdbacks, and reduced customer complaints.

The underlying cause of most of these issues is a lack of process standardization or a failure to follow approved processes. The key to improvement is to define these standards and create systems that ensure those processes will be followed correctly. An LPA program includes three critical things:

1. Auditors pulled from across the organization, including all levels of management
2. A set of simple audits, which do not require specialized knowledge to conduct, focused on high-risk processes
3. A system of reporting and follow-up that exposes problem areas and ensures containment and corrections are put – and held – in place.

In our modern world, we are finding out that *product inspections* are not enough. *Process audits* are an important, but often overlooked, as being part of a holistic quality system. Most standard quality systems are focused on product sampling and inspections – testing the actual output of a manufacturing or servicing process. This is necessary and important but doesn't go far enough. To be sure, non-conformances will likely be caught, but because of the delay and small sample sizes, significant rework and scrap costs may have already been incurred. Catching non-conformances earlier can significantly reduce COQ. In addition, some quality problems, especially those related to durability, cannot be caught with standard product inspections. A part may look fine leaving the plant, but if it was built incorrectly, will exhibit quality issues later. This can lead to significant costs, including customer complaints, warranty costs – even recalls.

So, process audits look at *how* products are made and services are rendered, and they can expose the non-conforming processes much earlier. Most process audits are conducted ad hoc or less formally, as in a *Gemba* walk. On the other hand, an LPA program introduces a systematic approach to process audits that improves both quality and performance. One may say that even the most rudimentary LPA is a process audit in steroids. Why? Because in an LPA program, auditors are drawn from the quality organization, but the team also includes individuals from cross-functional and multi-discipline areas of the organization, including executive leadership (these are the layers of LPA). Furthermore, in an LPA system, audits are conducted at a predetermined frequency depending on the level (or layer) of the auditor. Supervisors often conduct frequent process audits in their own area, while higher-level managers conduct the same audits less frequently and always over a broader range of areas. By conducting multiple layers of the same audit, an LPA system helps ensure that the audits are conducted accurately because the auditors are essentially double-checking each other. And by including multiple layers of management, the company demonstrates that quality is important to everyone.

The questions (checklists) that are part of an LPA are written to avoid the need for specialized knowledge because people from throughout the organization are involved in conducting the audits. In addition, it forces standardization in the pursuit of findings – whatever they are. To be sure, LPAs usually focus on processes and devices where deviation from the standard represents the highest risk of deviations or defects. However, this is NOT an exclusive practice. LPAs also look for processes that could be improved.

Here we must emphasize that LPA systems are only truly effective when they integrate *action, analysis, and improvements.* If an audit area does not pass, the auditor conducting the audit should record their finding and immediately take steps to prevent potentially defective products from getting out the door – containing the problem. Usually a corrective action can be anticipated and even taken immediately by the auditor. However, many corrective actions are more complex and involve other people and certainly more time. Also, corrective action must be for the specific "problem" at hand and at the "escape point," NOT on the "catching" spot. An escape point is the point where the non-conformance was created but was not caught. A catching spot, on the other hand, is the point where it was found.

The LPA system should account for this, track the assignment of the corrective action, and follow it through to completion – closing the loop on quality. Robust LPA systems capture all of this information and make it readily available for later analysis. A good system will highlight problem areas and help management recognize improvement opportunities. We will be amiss if we do not mention the second part of the corrective action. That is: *Prevention.* It is the action that the organization must take to avoid *a repeat* of the problem. If we do not take action with prevention, we may have an excellent corrective action but we will find ourselves "fixing" the same problem over and over again.

Now that we have given a general overview of what an LPA is, let us examine why this particular audit should be conducted. Whether we admit it or not, the fact is that organizations face increasing complexity and change, and are finding it challenging to account for the full COQ, including realized risk events, to the tune of tens of billions of dollars in costs across the major industries. From personal knowledge, one of the major OEMs reported a warranty cost for the year 2018, as 2.3 billion dollars. It is unfortunate that many organizations are missing a key opportunity to limit and control risk; they are not auditing their operations and processes as comprehensively as their products. In fact, many of them see audits as an additional cost to their business without any added value.

By looking beyond traditional product audits, which are reactive, and adopting process audits, which are inherently proactive, organizations can ensure that key activities are done properly and consistently, thereby improving quality at each and every stage of production. *LPAs prevent issues that product audits either can't identify or identify too late, requiring costly repair or scrapping, or even worse.* In addition, the best kept secret is that a solid LPA will – and does – identify error proofing systems that have become ineffective over time, highlight shortcomings before they become serious problems, and guide management on how to direct resources to review symptoms, identify problems, and find root causes. In other words, LPAs encourage continuous improvement. Most importantly, a well-implemented layered audit program will help instill a culture of quality within an organization, as senior level managers and executives actively demonstrate the importance of quality to the overall organization.

Undoubtedly then, the value of an LPA system is fundamentally measured in error reduction and improvement of product or service in the organization as well as customer satisfaction, which they appear in many ways, such as:

- *Rework costs:* including work that must be re-done because it was completed improperly in the first place. Because it occurs out of sequence, rework generally incurs significantly higher costs than normal operations. In fact, in no uncertain terms, rework adds additional processing and that is costly.
- *Direct material costs:* represent the raw materials used in production which have to be thrown away, due to overestimation in purchasing or even extended inventory to decrease unit cost.
- *Maintenance error costs:* although ongoing maintenance is required to keep delivering at the highest levels, errors here can increase production issues leading to scrap or more directly lead to costly acceleration of equipment failures. A good preventive maintenance (PM) can be helpful here. In addition, the overall equipment effectiveness (OEE) should be monitored to see if there is a problem in any of the three indicators, which are: availability, performance, and quality. (OEE is: availability $\times$ performance $\times$ quality and the best is 100%. Do not fall for the misconception of OEE being 80% as a satisfactory percentage. If the OEE is >100%, there is a cycle problem. It can NEVER be larger than 100%. OEE is fundamental metric in calculating capacity.)
- *Warranty costs:* these costs and similar reserves are used to pay for the defects that are identified post-delivery, typically initiated due to dissatisfaction by end customers.

- *Inspection costs:* these reflect the total cost to the organization for dedicated and non-dedicated resources assigning, executing, managing, and reviewing audit and inspection activities.

Implementing an LPA program can be challenging, but it also comes with significant rewards that include reduced quality costs and a more robust quality culture. Perhaps one of the best is to invest in a good software program that will do the record-keeping, analysis, and organization of the audit. It also expedites the retrieval process while providing a closed-loop process that ensures adherence to standards (https:// www.ease.io/best-practices-in-creating-a-layered-process-audit-program/). Perhaps the most important thing automotive manufacturers should know about LPAs is the need to *close the loop* on audit data (Stoop, 2017). That means:

- Addressing minor non-conformances on the spot
- Moving larger issues to a formal corrective action process
- Adding new audit questions to verify you're holding the gain.

Too many companies – particularly those overwhelmed by the details of LPAs – let audit data just sit there while quality issues continue driving up costs. Not only does ignoring problems increase consumer safety risks, it also jeopardizes the business itself. The old adage of "the boat keeps filling up with water because of leaks on the hull of the boat, and it will sink, unless the leaks are stopped" is quite appropriate here. The earlier we fix the leaks (small problems), the more successful we will be. LPAs have helped automotive suppliers achieve significant quality cost reductions and provide a structured framework that fosters a culture of quality. The trick is managing the complexity of these unique audits to find a good automation system. Software, in this case, will help companies finally get ahead, so the focus can be less on problems, more on prevention, and certainly more on the big, visionary goals of the future. In essence, a good software will expedite the process to produce results quickly, even with limited resources – for any organization. For example: a good automated LPA software built around the *Automotive Industry Action Group (AIAG) CQI-8 Layered Process Audits Guideline* allows organizations to:

- *Reduce administrative overhead:* An automatic system will reduce scheduling time and send communication alerts to individuals who need to know. Furthermore, it will facilitate the distribution and updating of checklists to the appropriate links of the team.
- *Improve resource efficiency:* When discussing efficiency that means "allocation of resources," which includes individual assignments as needed to value-added activities that meet the organization's strategic plan(s).
- *Get instant visibility into data:* The current proliferation of mobile electronics makes audit data available on time and to everyone who is authorized to see or review it. This immediate opportunity makes decision makers able to pinpoint issues/problems much faster as well as helping them in predictive quality indicators.

- *Increase audit completion rates:* Because the data is immediately available (access in real time), the completion of the audits is much faster and thus becomes a motivating factor for everyone in the organization to generate correct metrics as well as health charts to improve quality.
- *Protect data integrity:* Without question ALL data must have integrity. That means: that they have been (a) collected randomly, (b) without bias, (c) within the rules of frequency, and (d) kept with appropriate safeguards until used.
- *Refine checklists often:* It has been said many times that *the only thing for sure is that everything changes in due time.* Checklists fall in this category of changing. They must be changed to reflect the *dynamism* of the process. That change may be partial (modify questions or completely change or replace questions in their entirety) or complete (rewrite all questions). It must be remembered that by definition the LPA is an agile process and the key ingredient that makes it effective is precisely the focus on continual improvement. That means, change is expected!

Yet another way to maximize the benefits of the LPA is to focus on the traditional quality model of the **P**lan-**D**o-**C**heck-**A**ct (PDCA) approach:

- *Plan*: Learning about LPAs, creating your team, writing questions, and determining audit layers.
- *Do*: Implementing your LPA program requires disciplined scheduling, as well as a closed-loop system to manage findings.
- *Check*: Analyzing the data is key to assessing performance and leveraging results for further gains. (Deming (2000) prefers the term **S**tudy to indicate an in-depth analysis of the data).
- *Act*: Focusing on continuous improvement which implies ongoing engagement and updates to questions.

1. *Learn about LPA best practices*: Managers and process owners need a firm grasp of the LPA process before building teams and training participants. A good place to start is *the Automotive Industry Action Group's CQI-8 Layered Process Audits Guideline*, which provides detailed information on the following:
   a. How to explain LPAs to your team
   b. Types of checklist questions to include, and what to avoid
   c. Best practices for monitoring, measuring, and maintaining your LPA program.
2. *Create your LPA team*: Your audit management team is responsible for creating checklists and implementing the system across the organization. Experts recommend creating your team with employees from cross-functional areas and multi-disciplines of the organization, including executive members. (Whether or not you can realistically include a broad cross-section of the company may be a concern due to logistical or some other internal problems of availability. However, the point is: you shouldn't only

include experts and people of YOUR preference. If you do, they will tell you what you really want. It will not be effective. So, make sure that the team has "fresh eyes" as members and hopefully they will see what others have missed. Along with the fresh eye principle, try to rotate the auditors so that data integrity is kept and the "buddy-passing" is discouraged.

3. *Write your audit questions*: Each LPA should contain enough questions that would take 15 minutes to complete. It sounds simple, but in reality, getting fruitful answers depends on the questions you ask. Therefore, creating the questions must be thoughtful, direct, and concise. That takes time and preparation. A good source for generating thoughtful questions is historical data (from same or similar process(es) or the PFMEA or Process Flow Chart). A good audit question has three characteristics. They are:
   a. *Objective:* Questions need a clear answer, which is why most LPA questions are yes or no. This may seem "odd" since in most quality endeavors we ask with *open-ended* questions. The reason for a YES or a No expected answer is to make sure the auditee is comfortable in answering the question and also not to intimidate him/her. It is very important to establish a positive climate in the experience of asking questions and receiving the answer. We want to project confidence and trust the auditee. Be aware of body language as well.
   b. *Specific:* Avoid jargon, acronyms, and vague phrases. They can be misunderstood. Give the auditee full sentences with clear specifications, tolerances, or criteria for verification.
   c. *Concise:* Don't get bogged down in technical details. Use the KISS (Keep It Simple Stephanie) principle. The simpler the better. Make sure that people unfamiliar with the process are able to understand and answer the question.
4. *Create your audit plan:* LPAs get their name from the multiple layers of personnel who conduct the audits. To create the audit plan, the team must define the layers as well as the frequency for each layer. The most common hierarchy of levels is a three-level (layer) approach.
   a. *Layer 1* includes supervisors and team leads conducting audits daily across every shift.
   b. *Layer 2* includes middle management conducting audits once or twice monthly.
   c. *Layer 3* includes plant managers conducting audits quarterly, as well as executives conducting audits.
      Obviously, the frequency depends on the organization, process, and or historical level of issues/problems (both internal and customer).
5. *Implement the LPA program:* Once you've completed the planning process, it's time to execute the plans. You may have the best plan, but if you do not execute it, it is useless. The implementation process has to be realistic in reference to timing and availability of resources. Essential elements of a strong rollout include, but are not limited to
   a. *Efficient scheduling and follow-up:* LPA programs can require many audits annually. That is why we recommended earlier a software that

reduces the resulting administrative burden allowing management to assign audits and set up auto-notifications in minutes.

b. *Communication:* Also critical to implementation is communicating the benefits and expectations around the LPA program. Emphasize that LPAs aren't about blame – the goal is working together to (a) identify problems, (b) correct problems, (c) identify preventive approaches to the identified problems, (d) create higher quality, (e) create leaner operations, and (f) focus on continuous improvement.

c. *Corrective action:* Unless you have a closed-loop process in place to manage non-conformances, the rest of your LPA efforts are pointless. You should be able to quickly close out smaller non-conformances and assign corrective actions on the spot. However, do not forget to identify preventive methods for avoiding the same problem in the future. If you stop at the correction stage, that means eventually you going to repeat the failure.

d. *Measure and improve your LPAs:* The final step in implementing an LPA program is analyzing your data and taking action based on your findings. The basic question that has to be asked is: *are you better off now or before the implementation of the LPA*. If the answer is NO, something went wrong or the process of the LPA was not followed correctly. If YES, then it should serve as a motivator to do it again and again. So, it is very important to measure the LPA program itself so that, leading metrics such as audit completion rates and time to closure for corrective actions may be developed. Focus on steps such as

   i. Reviewing audit data while it's still actionable instead of weeks or months later (which is often the case with paper checklists and spreadsheet-based tracking). Not reviewing the LPA results is a common occurrence and the reason why management feels that they are not value added.

   ii. Updating questions when processes or requirements change, as well as to verify corrective actions are working. This is another issue with the LPA. Many do not take the time to update the questions and obviously, by their own doing, the results are not accurate or timely.

Compared to traditional inspections and audits, LPAs provide added layers of security to catch process variations before they cause defects or safety incidents. By generating large volumes of data, LPAs also increase visibility into the drivers of quality costs. In our experience, metrics where LPAs have the biggest impact include the following:

* *Scrap and rework*: Companies that implement LPAs have been able to cut scrap costs in half over the course of several months.
* *Customer returns*: LPA implementation has been tied to a more than 50% reduction in customer defects in some organizations. Other companies have reported going months or longer without any customer returns.

- *Total cost of quality*: Reducing quality costs has a direct impact on bottom-line revenue growth. One reason is that quality costs increase exponentially as you move from prevention to detection to correction of actual failures.

Beyond the financial benefits, LPAs also help companies develop a *culture of quality*. When fully implemented, LPA programs enable

- *Company-wide engagement*: LPAs break quality out of its administrative silo, involving people from all departments including executives.
- *Proactive risk management*: LPAs identify issues that impact customer satisfaction upstream, rather than at the point of manufacture. Their structure also supports the central goal of corrective action, which is permanently reducing risk, if prevention is incorporated into the correction practice.
- *Top-level prioritization*: Making management a visible presence on the plant floor is powerful evidence of an authentic commitment to quality. The frequency of LPAs is a signal to everyone that quality is the business of everyone and as such ALL employees should focus on proactive improvement.
- *Ongoing communication*: LPAs give operators the chance to share observations and suggestions directly with management. Ongoing communications are a way to recognize ownership of process, responsibility of the actions taken, and a pride for what is being done. When people see leadership cares enough to be present and fix problems, they're more likely to speak up.
- *Accountability*: LPAs generate much data, but it is up to management to see that are taken seriously and implement the appropriate and applicable corrective actions. In addition to finding out week spots in the process, it gives management an opportunity to identify individuals for special recognition.

It must be emphasized here that cultural benefits aren't always as tangible, but they're essential building blocks for achieving any ambitious quality cost reductions.

CHALLENGES OF LAYERED PROCESS AUDIT IMPLEMENTATION

While many manufacturers may see the benefits of LPAs, there are just as many who struggle with its implementation due to many reasons including inefficiencies in the paper processing of data (tracking), questions, and so on. Fundamentally, however, there are five specific issues that cause "headaches" and challenges for the implementers. They are as follows:

- *Inefficient scheduling*: A given plant may need to conduct many audits however, the resources are not available to carry out the scheduling requirements. As a result, the motivation and the enthusiasm is lost and quite often never recover.
- *Low audit completion rates*: Organizations often struggle to keep up with the high frequency of audits required, primarily because of change in priorities. When the organization does not stay on plan, the probability is very low that the LPAs will be completed.

- *Data integrity problems*: All studies are as good as the data collected. LPAs are no different. Collecting good data from LPAs is impossible if people are not convinced that everyone is involved in this *quality culture of improvement*. What happens when they are not convinced? They go through the motions of conducting an audit without thought, conviction, and more importantly without respect for the information they collect. Needless to say, the audit for all intents and purposes is worthless.
- *Analytical bottlenecks*: Bottlenecks must be the priority of any good planned LPA. Eliminating that challenge will improve productivity and reduce inventory stock to accommodate any shortages in the next process following the bottleneck operation. If this evaluation is manual, the chances of conducting a thorough analysis are slim due to the high volume of information. This is another reason why a good software may help. Bottleneck operations can cause companies to compromise the deadlines for delivery, not finding problems until too late and produce defective product. This in turn generates a risk to companies that a bad "non-conforming" product will be delivered to the customer.
- *Outdated checklists*: The blood life of any audit, including the LPAs, are questions that are relevant and current of the process being audited. Therefore, since processes change it is imperative to review and change as necessary the questions that are planned for the audit. There is nothing worse than recycled questions that are either outdated or plain unacceptable. So, before the audit commences, take some time and review and update as required by the checklist to account for changing risks and lessons learned. Recalling old checklists and distributing new ones is one of the first tasks to go when people get busy. Here we must sound the alarm. Many standards and customer specifications have their own predetermined checklists that are required to be followed (*e.g.* ISO 9001, ISO 14001, IATF 16949, AIAGs FMEA, AIAG/VDA FMEA and process audit VDA 6.3). Whereas these check lists are very good for initial audits, we are concerned about the long-term effects on their efficacy and effectiveness. If the supplier knows ahead of time what the questions are and how they are expected to answer, we doubt very much if in the long term these questions are appropriate and applicable – if we assume continual improvement.

The problem for manufacturers is that LPAs aren't just about checking some boxes – although many organizations feel and act that they are. How unfortunate! There's no silver bullet to guarantee success. However, at least in the automotive industry – what matters is to always remember that the customer wants your organization to drive continuous improvement and be able to prove it. The best way to do that is through a systematic LPA of your processes.

ERROR PROOFING AUDITS

Continual improvement is a crucial part of any Quality Management System (QMS), and *IATF 16949* brings it to the next level by requiring problem-solving and error proofing processes as part of the requirements for continual improvement. Error proofing represents a structured approach to ensuring the quality of products throughout the entire manufacturing process. It provides organizations with a tool to improve the manufacturing or business processes to prevent specific errors – and, thus, defects – from occurring. Error proofing methods enable organizations to discover sources of errors through fact-based problem-solving. The focus of error proofing is not on identifying and counting defects; rather, it is on the elimination of their cause: one or more errors that occur somewhere in the production process (Stojanovic, 2017).

As important the error proofing audit is, we must not become complacent to the methodology as a panacea to all our problems. The reason is that even though error proofing is a solid methodology, it is not 100% guarantee that the problem will be caught or prevented. The only method of error proofing that guarantees 100% prevention is the specific solution to a particular problem with *orientation* issues. All others can fail for many reasons, including degradation over time, limit switches may break their antenna signals, paper stickers may lose their glue, photo cells may become nests for spiders, and so on.

What Is an Error?

Errors are inadvertent, unintentional, accidental mistakes made by people because of the human sensitivity designed into our products, which result in those "once in a while" defects that we always find difficult to control. Inadvertent errors are not only possible, but inevitable, and they happen regularly. It is important here NOT to blame the operators. They are not the problem, unless there is an issue of sabotage. Most of the problems are process problems with a heavy responsibility assigned to the design. As humans we will never design perfect products. However, that does not mean that we should stop designing new products. So, even though we know of the imperfect designs, we can circumvent these flaws with building controls and tests to catch them before they reach the customer.

Mistakes, or errors, are part of our everyday lives. It is hard to avoid them, and even harder to explain why they happened. Being an ever-present part of our lives, it is hard to avoid errors in the workplace as well. Examples of work-related errors are as follows:

- Missing parts – forgetting to assemble a part: screws, labels, orifice tubes ...
- Misassembled parts – incorrect assembly: loose parts, upside down, not aligned, *e.g.* brackets (backwards), seals (not aligned), screws (loose), labels (upside down) ...
- Incorrect processing – disposing of a part rejected during testing in the wrong pile
- Incorrect parts – retrieving and assembling the wrong part from a model mix selection: seals, labels, brackets, cases ...

And, of course, these examples are related to the production process mainly, but there are many more in different situations. Depending on the type of error, the stakes can be high. Work-related errors can lead to work-related injuries, loss of money, wasted time, loss of jobs or deals, and even injury or death of consumers, so it is no wonder that this is an important part of a standard that focuses on quality in the automotive industry. So, a fertile ground for error proofing audits are the areas which depend on a heavy presence of human inspections and deal in confined areas of work, safety issues, and environmental concerns.

WHAT DOES THE STANDARD REQUIRE?

Requirements for error proofing processes are part of the continual improvement requirements, and an entire subclause (ISO 9001:2015 and IATF 16949: 2016; 10.2.4) is dedicated to them. The standard requires an organization to establish a documented process to determine the use of appropriate error proofing methodology. However, the methodology for error proofing should be documented in the process risk analysis, such as PFMEA (Process Failure Mode Effect Analysis), and the test frequency should be documented in the control plan. The process must include testing of the error-proofing devices for failure or simulated failure, and the error proofing device failures should have a reaction plan. And finally, of course, records of this process should be maintained.

HOW DO WE ERROR PROOF?

Error proofing is the activity of awareness, detection, and prevention of errors that could adversely affect our customers (with defects) and our people (with injuries), and result in WASTE! (see Figure 7.6). Therefore, let us define these with little more context:

- *Awareness*: Having the forethought that a mistake can be made, communicating the potential, and planning the design of the product or process to detect or prevent it.
- *Detection*: Allowing the mistake to happen but providing some means of detecting it and alerting someone so that we fix it before sending it to our customer.
- *Prevention*: Not allowing the possibility for the mistake to occur in the first place. For this, there are two main techniques for error proofing:
  1. *Design for Manufacturability (DFM)*: This is a technique that results in designs that cannot be incorrectly manufactured or assembled. This technique can also be used to "simplify" the design and therefore reduce its cost.
  2. *"Poka-Yoke" system*: Devices or inspection techniques are put in place that assure that setup is done correctly; *i.e.* 100% good parts are produced from the first piece onward. It is important to note that more often than not, regardless of the technique you decide to apply (and you will probably decide to apply them both), the error proofing process has some common steps that ensure success. They are as follows:
     - *Identify*: The organization needs to identify error proofing opportunities through PFMEA, quality data, warranty data, brainstorming

(questions to ask, freeform ...), *etc.* The best way to identify these opportunities is through an active team discussion and team brainstorming.

– *Analyze*: The organization needs to prioritize the error-proofing opportunities, determine the level of error-proofing required, brainstorm, and select the error-proofing mechanisms. Always remember that NOT everything is important. Force the team to prioritize. *Force field* analysis is a good method to do precisely this activity.

– *Plan*: The organization needs to determine process mechanisms for error proofing, define actions plans, and create the error-proofing control plan. Without a predefined plan, everyone is going to do what they think is important and efficient. The loss of the objective will be lost.

– *Implement*: The organization needs to implement various activities to put the error-proofing process into practice, such as installation, validation, check sheet/log, and operator instructions. As we already have mentioned it, even if there is an exceptional plan in the record, if it is not implemented, there is NO VALUE to it. So, implement and you may be pleasantly surprised when it works. If not, start planning again, using the PDCA model.

– *Evaluate*: Finally, the organization needs to evaluate the effects of the actions taken, and if the number of errors has not decreased as expected, an additional set of actions should be taken. Evaluation using objective criteria is a must, in deciding the effectiveness of the error proofing audit.

CORRECTIVE ACTION AUDITS

Clauses 10.2.1–10.3.1 of ISO 9001:2016 and IATF 16949: 2018 require the organization to assess the effectiveness of the QMS, and that system includes a process for handling corrective actions as per documented procedure.

An audit of that process would therefore begin with a review of the documented procedure and include a sample of corrective actions performed and recorded in order to determine whether the process is effectively implemented in order to prevent recurrence of non-conformities. Subclauses a through f should be confirmed to be included in the process. Corrective actions (clause 10.2.1) are referenced in internal audits (9.2.2), but the process might also be used to address customer complaints and/or other identified issues judged important enough to address in a controlled manner.

The cost of ineffective corrective action can be astronomical when you consider the monetary and reputational impact of delayed problem-solving. On a small scale, repeat problems – even minor errors – send a message to customers that you just don't care to get it right. The interesting follow-up of this is that the *laissez faire* attitude leads to more significant quality problems and with it comes loss of integrity and trust. Therefore, to ensure corrective actions reduce risk, automotive suppliers need to avoid key mistakes around measuring effectiveness, root cause analysis, and tracking closure (Faircloth, 2018).

1. *Not measuring effectiveness over time*: Unless you measure something, there is no other way to know where you are and how much change has occurred. Most often, corrective actions fail because companies fail to measure their effectiveness over time. This is true for individual corrective actions as well as for the corrective action process as a whole.

   After a corrective action is complete, you need a way to determine not only whether it was effective but also whether you're holding the gain. LPAs are one of the best methodologies to use for tracking, since as part of their evaluation of "the" process not only verification of actions has taken place but also validates possible high-risk areas and opportunities for further improvement. Typical monitoring metrics may be
   * Total number of corrective action requests as opposed to completed.
   * Number of overdue requests. Days overdue for closure (*e.g.* 30, 45, and/ or 60 days overdue. The most typical is 45 days. Of course, this depends on the complexity and availability of resources).
   * Average time to closure (see below, #4).
2. *Shallow root cause analysis*: Another major barrier to continuous improvement is ineffective or "check the box" approaches to root cause analysis. Frequently you have someone in an office doing the analysis, blaming every problem on operator error with training as the easiest solution. Instead, organizations need to make problem-solving a team effort that actually involves people on the shop floor with first-hand knowledge of the issue at hand. Just as important, you need to be willing to ask hard questions and take an honest look at the answers. One way to get to the root cause is to use the "5 Whys" analysis, in which the management asks "Why?" after an answer regarding the root cause to an issue. This interrogative technique is used to fully see the cause-and-effect relationships underlying a problem. It's only when you see the full picture that you get to the true root of quality issues. The goal of any good root cause analysis is to find the "escape point of the problem." Unless this happens, the problem will occur again and again.
3. *Lessons aren't shared*: The third reason why corrective actions fall short is failure to translate lessons learned from corrective action to other sites, products, and processes. Everyone talks about Things Gone Wrong (TGW) and Things Gone Right (TGR), but not too many address a thorough review to take advantage of what was learned. Obviously TGW have to be addressed and fixed – as soon as possible. However, TGR must also be reviewed so that they can be repeated again. (It may be cheaper to skip the follow-through, at least in the short run. But ultimately, the price of letting problems become systemic is far steeper, impacting your most important client relationships and even the business as a whole.)
4. *Long closure times*: Taking too long to fix problems is not something regulators tolerate, imposing millions in fines to individual companies for failing to fix problems in a timely manner. While it's true that some complex issues can take weeks (or longer) to correct, it's critical to recognize that delayed closure times increase the risk of quality escapes. In fact, average time to closure for corrective actions is often a leading indicator of quality

problems. Leading indicators are metrics that point to what might happen in the future, compared to lagging indicators that measure results.

5. *Administrative pitfalls*: Many organizations still use manual or paper-based approaches to managing corrective action, which creates many opportunities for requests to get delayed. Filling out paper forms, assigning corrective actions, following up with responsible individuals – miss any one of these tasks and your corrective action process falls apart. Automation (through software use) can help eliminate the cracks in your process, providing additional benefits for those who integrate corrective action with internal process audits. A typical platform allows you to

- Assign mitigations and corrective actions on the spot during mobile audits to promote problem-solving culture
- Make corrective action a team effort by routing requests to the right people at the right time
- Shorten time to closure by escalating missed steps to supervisors
- Add new questions to audits based on past corrective actions for continuous verification of effectiveness.

Closing the loop through automation can eliminate many of the problems discussed here and set the foundation for a *positive culture of quality*. An authentic commitment to solving problems is at the heart of quality culture, making corrective action a high-impact area where small changes can lead to big results.

DISTANCE AUDIT

In the last five or so years, distance audits are proliferating in the field of quality. I am fully receptive to performing a substantial portion of the audit remotely as opposed to a client's site because I believe both can promote a similar level of world-class client service. However, there is resistance by some for a variety of reasons, which I also understand. To fully embrace remote auditing, we must maintain the crucial relational aspect and the learning opportunities we experience through in-person interactions. This is very difficult to accomplish and it will take some time before we have the data to support 100% distance auditing.

For years we all have been conditioned to perform audits at the GEMBA location or as close to the Gemba as possible (Imai, 2012). It is possible with modern technology to do a portion of the audit in distance format but not the entire audit. For example, one may do the "desk audit" – traditionally held in the supplier's conference room – via a distance technology. No problem. However, when it comes to the actual physical audit, there are issues that may concern everyone. To be sure, modern technology provides tools to compensate for some, but we all know that the way a picture or a video is taken (angle, perspective, depth of field, zoom or not, and so on) can play a role in deciding the results of an audit. So, for document review the distance audit is acceptable. However, for the physical audit at the supplier's facilities, the jury is still out. In the final analysis, the issue is resolved when both you and the client are comfortable with the concept of conducting the audit 100% remotely. By doing so, the following are expected:

- Remote auditing significantly increases quality of life because it allows you to work from anywhere.
- No one wants an auditor in their building taking up valuable conference room space. It creates tension and uneasiness.
- You will never be told that a step ladder and a one-foot fold-out tray is the only viable workspace again.
- You don't need to drive 2 hours each way to your client's site.
- Remote auditing creates more focus and transparency, especially when using specialized tools such as Box, Microsoft Teams and others.

There is no doubt that there are challenges in conducting audits 100% remotely. If that is the plan, the following should be addressed:

- The absence of evaluating responses in relationship to the body language is of paramount importance. Seeing people's reactions to questions and dialog provides valuable information to aid in performing an audit.
- When parties don't treat Skype calls or Zoom Skype with the same attention as a face-to-face meeting, they can get distracted.
- Remote auditing requires the client to appreciate the use of technology and encourages them to desire a change in their habits.
- Technology systems need to be upgraded across industries to capture more data electronically.

INTERNAL SYSTEM AUDIT

The internal audit system is the sum of all policies and procedures arranged by the organs of the company (supervisory board/board of directors/management board) that guarantee an orderly and efficient management, the protection of assets, the prevention or detection of tortious acts and errors, the accuracy and completeness of the records of management accounting, and the timely preparation of reliable financial information. Companies are required to operate an internal audit system by numerous legal requirements, *e.g.* KonTraG (the EU version of the SOX), Sarbanes-Oxley Act (SOX) as well as, international standard and industrial standards as well as organizational requirements. Due to the large amount of records and documents, the internal audit system cannot be administered by the controller of the organization. In fact, due to their complexity, each organization has their own audits for their particular departments. In quality specifically, the internal system audit is called an internal audit or a first-party audit and generally focuses on the following areas, policies, and practices:

- Business Conduct Standards
- Internal Quality Audits
- Risk Assessments
- Internal Controls
- Internal Facility Projects
- Method Validation Audits
- An Environment, Health, and Safety (EHS) Audits.

Epilogue

THE THREE DIFFERENT TYPES OF AUDITS

ISO 19011:2018 defines an audit as a

> systematic, independent and documented process for obtaining audit evidence [records, statements of fact or other information which are relevant and verifiable] and evaluating it objectively to determine the extent to which the audit criteria [a set of policies, procedures or requirements] are fulfilled.

There are several types of audits; however, the primary ones are of three types.

PROCESS AUDIT

This type of audit verifies that processes are working within established limits. It evaluates an operation or method against predetermined instructions or standards to measure conformance to these standards and the effectiveness of the instructions. A process audit may

- Check conformance to defined requirements such as time, accuracy, temperature, pressure, composition, responsiveness, amperage, and component mixture.
- Examine the resources (equipment, materials, people) applied to transform the inputs into outputs, the environment, the methods (procedures, instructions) followed, and the measures collected to determine process performance.
- Check the adequacy and effectiveness of the process controls established by procedures, work instructions, flowcharts, and training and process specifications.

PRODUCT AUDIT

This type of audit is an examination of a particular product or service, such as hardware, processed material, or software, to evaluate whether it conforms to requirements (*i.e.* specifications, performance standards, and customer requirements).

SYSTEM AUDIT

An audit conducted on a management system. It can be described as a documented activity performed to verify, by examination and evaluation of objective evidence, that applicable elements of the system are appropriate and effective and have been developed, documented, and implemented in accordance and in conjunction with specified requirements.

- A *quality management system audit* evaluates an existing quality management program to determine its conformance to company policies, contract commitments, and regulatory requirements.
- Similarly, an *environmental system audit* examines an environmental management system, a *food safety system audit* examines a food safety management system, and *safety system audits* examine the safety management system.

Other methods, such as a desk or document review audit, error-proofing audit, distance audit, and others, may be employed independently or in support of the three general types of audits. Some audits are named according to their purpose or scope. The scope of a department or function audit is a particular department or function. The purpose of a management audit relates to management interests, such as assessment of area performance or efficiency.

An audit may also be classified as internal or external, depending on the interrelationships among participants. Internal audits are performed by employees of your organization. External audits are performed by an outside agent. Internal audits are often referred to as first-party audits, while external audits can be either second party or third party. More detailed information, see https://asq.org/quality-resources/auditing. Here we summarize the function of each:

- A *first-party audit* is performed within an organization to measure its strengths and weaknesses against its own procedures or methods and/or against external standards adopted by (voluntary) or imposed on (mandatory) the organization. A first-party audit is an internal audit conducted by auditors who are employed by the organization being audited but who have no vested interest in the audit results of the area being audited.
- A *second-party audit* is an external audit performed on a supplier by a customer or by a contracted organization on behalf of a customer. A contract is in place, and the goods or services are being, or will be, delivered. Second-party audits are subject to the rules of contract law, as they are providing contractual direction from the customer to the supplier. Second-party audits tend to be more formal than first-party audits because audit results could influence the customer's purchasing decisions.
- A *third-party audit* is performed by an audit organization independent of the customer–supplier relationship and is free of any conflict of interest. Independence of the audit organization is a key component of a third-party audit. Third-party audits may result in certification, registration, recognition, an award, license approval, a citation, a fine, or a penalty issued by the third-party organization or an interested party.

In the case of the internal audits, there is no certification. The end result is to identify and correct non-conformances and suggest improvement opportunities. Whether or not those activities take place depends on the management of the organization, and the intensity of implementing actions depends on the attitude of the entire organization in terms of "culture quality".

In the case of the second-party audit, even though there is no certifications, there is an approval of the customer as to the effectiveness of the organization's QMS.

In the case of the third-party audit, there is a certification. However, that certification is given by organizations that have been evaluated and accredited by an established "accreditation board," such as the ANSI-ASQ National Accreditation Board (ANAB).

No matter what kind of audit an organization is participating and is held accountable for, all of them to some degree are interested in "the" value added that the audit will generate to the organization. So, value-added assessments, management audits, added value auditing, and continual improvement assessments are terms used to describe an audit purpose beyond compliance and conformance. The purpose of these audits relates to organization performance. Audits that determine compliance and conformance are not focused on good or poor performance, yet. Performance is an important concern for most organizations.

A key difference between compliance audits, conformance audits, and improvement audits is the collection of evidence related to organization performance versus evidence to verify conformance or compliance to a standard or procedure. An organization may conform to its procedures for taking orders, but if every order is subsequently changed two or three times, management may have cause for concern and want to rectify the inefficiency.

A product, process, or system audit may have findings that require correction and corrective action. Since most corrective actions cannot be performed at the time of the audit, the audit program manager may require a follow-up audit to verify that corrections were made and corrective actions were taken. Due to the high cost of a single-purpose follow-up audit, it is normally combined with the next scheduled audit of the area. However, this decision should be based on the importance and risk of the finding.

An organization may also conduct follow-up audits to verify preventive actions were taken as a result of performance issues that may be reported as opportunities for improvement. Other times organizations may forward identified performance issues to management for follow-up. So, in general terms, the cycle of all auditing may be summarized in four steps. They are as follows:

1. *Audit planning and preparation*: Audit preparation consists of planning everything that is done in advance by interested parties, such as the auditor, the lead auditor, the client, and the audit program manager, to ensure that the audit complies with the client's objective. This stage of an audit begins with the decision to conduct the audit and ends when the audit itself begins.

2. *Audit execution*: The execution phase of an audit is often called the *fieldwork*. It is the data-gathering portion of the audit and covers the time period from arrival at the audit location up to the exit meeting. It consists of multiple activities including on-site audit management, meeting with the auditee, understanding the process and system controls, and verifying that these controls work, communicating among team members, and communicating with the auditee.

3. *Audit reporting*: The purpose of the audit report is to communicate the results of the investigation. The report should provide correct and clear data that will be effective as a management aid in addressing important organizational issues. The audit process may end when the report is issued by the lead auditor or after follow-up actions are completed.

4. *Audit follow-up and closure*: According to ISO 19011, clause 6.6, "The audit is completed when all the planned audit activities have been carried out, or otherwise agreed with the audit client." Clause 6.7 of ISO 19011 continues by stating that verification of follow-up actions may be part of a subsequent audit.

   a. *Corrective action* is action taken to eliminate the causes of an existing non-conformity, defect, or other undesirable situation in order to prevent recurrence (reactive). Corrective action is about eliminating the causes of problems and not just following a series of problem-solving steps.

   b. *Preventive action* is action taken to eliminate the causes of a potential non-conformity, defect, or other undesirable situation in order to prevent occurrence (proactive).

References

Ad Hoc Committee on Professional Ethics. (1983). "Ethical Guidelines for Statistical Practice: Report of the Ad Hoc Committee on Professional Ethics." *American Statistician* 37, pp. 5–8.

AIAG. (2005). *Statistical Process Control, SPC.* 2nd ed. Southfield, MI: Automotive Industry Action Group.

AIAG. (2008a). *Advanced Product Quality Planning, APQP.* 2nd ed. Southfield, MI: Automotive Industry Action Group.

AIAG. (2008b). *Failure Mode and Effect Analysis, FMEA, FMEA.* 4th ed. Southfield, MI: Automotive Industry Action Group.

AIAG. (2009). *Production Part Approval Process, PPAP.* 4th ed. Southfield, MI: Automotive Industry Action Group.

AIAG. (2010). *Measurement Systems Analysis, MSA.* 4th ed. Southfield, MI: Automotive Industry Action Group.

AIAG/VDA. (2019). *Failure Mode and Effect Analysis – FMEA: Design FMEA and Process FMEA Handbook.* 1st ed. Southfield, MI: Automotive Industry Action Group/Verband der Automobilindustrie.

American Society for Quality Control. (1993). *Certification Program for Auditors of Quality Systems.* Milwaukee, WI: ASQC.

Anderson, B. and T. Fagerhaug. (2000). *Root Cause Analysis.* Milwaukee, WI: Quality Press.

ANSI/ISO/ASQC A8402. (1994). *Quality Vocabulary.* Milwaukee, WI: ASQC.

AP. (June 2, 2020). "Hydroxychloroquine Study Questioned as Journal Issues 'Expression of Concern' about Research." https://www.nbcnews.com/health/health-news/hydroxychloroquine-study-questioned-journal-issues-expression-concern-about-research-n1222636. Retrieved on June 7, 2020.

Arter, D. (1994). *Quality Audits for Improved Performance.* 2nd ed. Milwaukee, WI: Quality Press.

Arter, D. (2003). *Quality Audits for Improved Performance.* 3rd ed. Milwaukee, WI: ASQ Quality Press.

Bicheno, J. (2004). *New Lean Toolbox: Towards Fast, Flexible Flow.* Buckingham: PICSIE.

Birch, A. J. (1990). "Deceit in Science: Does It Really Matter?" *Interdisciplinary Science Reviews* 15, pp. 335–343.

Breyfogle III, F. (2003). *Implementing six sigma.* 2nd ed. New York: J. Wiley and Sons, Inc.

Brumm, E. K. (1995). *Managing Records for ISO 9000 Compliance.* Milwaukee, WI: Quality Press.

Brush, S. G. (1974). "Should the History of Science Be Rated X?" *Science* 183, pp. 1164–1172.

Burdick, R., C. Borror and D. Montgomery. (2005). *Design and Analysis of Gauge R&R Studies: Making Decisions with Confidence Intervals in Random and Mixed ANOVA Models.* Philadelphia, PA: SIAM.

Cartin, T. (1999). *Principles and Practices of Organizational Performance Excellence.* 2n ed. Milwaukee, WI: Quality Press.Bottom of Form.

Cartin, T. (2010). *Principles and Practices of Organizational Performance Excellence.* New Delhi, India: New Age International Publisher.

Carvalho, A. and P. Sampaio. (April, 2020). "A Feather in Your Cap." *Quality Progress* 53(4), pp. 42–49.

Clements, R., Sidor, S. S. and R. E. Winters. (1995). *Preparing Your Company for QS-9000; A Guide for the Automotive Industry.* Milwaukee, WI: Quality Press.

Cottman, R. J. (1993). *A Guidebook to ISO 9000 and ANSI/ASQC Q90*. Milwaukee, WI: Quality Press.

Day, R. (1993). *Quality Function Deployment: Linking a Company with Its Customers*. Milwaukee, WI: Quality Press.

De Carvalho, A. and P. Sampalo. (2020). "How Performance Excellence Models Are Doing 30 Years Later." *Quality Progress* 53(4), pp. 42–49.

Deming, W.E. (2000). *Out of the Crisis*. Cambridge, MA: The MIT Press.

Desilva, J. (May 5, 2020). "The Decline in ISO 9001 Certification: Does Quality Matter Anymore?" *Quality Digest*. https://www.qualitydigest.com/inside/management-article/decline-iso-9001-certification-does-quality-matter-anymore-050520.html

Diaconis, P. (1985). "Theories of Data Analysis: From Magical Thinking through Classical Statistics." In *Exploring Data Tables, Trends and Shapes*, eds. D. C. Hoaglin, F. Mosteller, and J. W. Tukey. New York: Wiley, pp. 1–36.

Duhan, S. (1979). "An Audit Is More Than an Audit." *ASQC Technical Conference Transactions*, pp. 94–108.

Dunnette, M. D. (1966). "Fads, Fashions and Folderol in Psychology." *American Psychologist* 21, pp. 343–352.

Eckes, G. (2001). *Making Six Sigma Last*. New York: J. Wiley and Sons, Inc.

Emmons, S. (1977). "Auditing for Profit and Productivity." *ASQC Technical Conference Transactions*, Philadelphia, PA, pp. 206–212.

Everett, R. and Sohal, A. (1991). "Individual Involvement and Intervention in Quality Improvement Programmes: Using the Andon System." *International Journal of Quality & Reliability Management* 8(2). doi: 10.1108/EUM0000000001635

Faircloth, S. (March 26, 2018). "5 Reasons Why Corrective Actions Miss the Mark." https://www.ease.io/5-reasons-why-corrective-actions-miss-the-mark/. Retrieved on July 15, 2020.

Fonseca, L. and P. Domingues. (October, 2017). "The Results Are In." *Quality Progress* 50(10), pp. 26–33.

Ford. (December, 2011). *FMEA Handbook Version 4.2*. Dearborn, MI: Ford Motor Company.

Friedlander, F. (1964). "Type I and Type II Bias." *American Psychologist* 19, pp. 198–194.

Galsworth, G. D. (2005). *Visual Workplace: Visual Thinking*. Portland, OR: Visual-Lean Enterprise Press.

Gapp, R., R. Fisher and K. Kobayashi. (2008). "Implementing 5S within a Japanese Context: An Integrated Management System." *Management Decision* 46(4), pp. 565–579.

Grossman, C. (January – February, 1995). "ISO 9000 Readiness Survey." *Quality in Manufacturing. No pages available.*

Guasch, J., J. Racine, I. Sánchez and M. Diop. (2007). *Quality Systems and Standards for a Competitive Edge. Directions in Development; Trade*. Washington, DC: World Bank. https://openknowledge.worldbank.org/handle/10986/6768 License: CC BY 3.0 IGO.

Hall, E. (1998). *Managing Risk*. New York: Addison-Wesley.

Hammer, M. and J. Champy. (1993). *Reengineering the Corporation*. New York: Harper Business.

Harry, M. and R. Schroeder. (2000). Six *Sigma: The Breakthrough Management Strategy Revolutionizing the World's Top Corporations*. New York: Doubleday.

Hirano, H. (ed.). (1988). *JIT Factory Revolution: A Pictorial Guide to Factory Design of the Future*. Cambridge, MA: Productivity Press.

Hirano, H. (1995). *5 Pillars of the Visual Workplace*. Cambridge, MA: Productivity Press.

Howard, P. (December 5, 2019). "Ford Workers Break Their Silence on Faulty Transmissions: 'My Hands Are Dirty. I Feel Horrible.'" *Detroit Free Press*. https://www.usatoday.com/story/money/cars/2019/12/05/ford-focus-fiesta-dps-6-transmission-problems/2617601001/. Retrieved on December 5, 2019. See also: https://finance.yahoo.com/news/ford-workers-break-silence-faulty-145800004.html. Retrieved on December 5, 2019.

Hutchins, A. (1992). *Standard Manual of Quality Auditing: A Step By Step Workbook with Procedures and Checklists*. Milwaukee, WI: Quality Press.

Imai, M. (2012). *Gemba Kaizen: A Commonsense Approach to a Continuous Improvement Strategy*. 2nd ed. New York: McGraw-Hill.

Irwin Professional Publishing. (June, 1993). "ISO 9000 Survey." Quality Systems Update. Burr Ridge, IL: Irwin Professional Publishing.

Isidore, C. (February 4, 2015). "GM's Total Recall Cost: $4.1 Billion" https://money.cnn.com/2015/02/04/news/companies/gm-earnings-recall-costs/. Retrieved on May 6, 2020.

ISO 9001:2015. (November 2018). *Quality Management System*. Geneva, Switzerland: International Organization for Standardization.

ISO 9001:2018. *Quality Management System - Model for Quality Assurance in Design, Development*. Zurich, Switzerland: International Organization for Standardization.

ISO 9001:2015. *Quality Systems—Model for Quality Assurance in Design, Development, Production Installation and Servicing*. Milwaukee, WI: ASQC.

ISO 9004:2018. *Quality Management and Quality System Elements—Guidelines*. Milwaukee, WI: ASQC.

ISO 14001:2015. *Environmental Management System*. Zurich, Switzerland: International Organization for Standardization.

ISO 18001:2007. *Health & Safety Management Standard*. Zurich, Switzerland: International Organization for Standardization.

ISO 45001:2018. *Occupational Health and Safety*. Zurich, Switzerland: International Organization for Standardization.

Johnson, B. (December 19, 2019). "GM Recalls 814K Vehicles to Fix Brake, Battery Problems." https://www.mlive.com/news/ann-arbor/2019/12/gm-recalls-814k-vehicles-to-fix-brake-battery-problems.html. Retrieved on May 4, 2020.

Johnson, M. (May 21, 2020). "CDC Acknowledges Mixing Up Coronavirus Testing Data." https://thehill.com/policy/healthcare/499085-cdc-acknowledges-mixing-up-coronavirus-testing-data. Retrieved on June 7, 2020.

Johnston, K. (December 19, 2019). "GM Recalls 814K Cars, Pickups Due To Brake, Battery Problems." https://baltimore.cbslocal.com/video/4258876-gm-recalls-814k-cars-pickups-due-to-brake-battery-problems/. Retrieved on December 20, 2019).

Keeney, K. A. (1995a). "The ISO 9000." Auditor's Companion. Milwaukee, WI: Quality Press.

Keeney, K. A. (1995b). *The Audit Kit*. Milwaukee, WI: Quality Press.

Krisher, T. (June 13, 2020). "Ford Recalls about 2.5 Million Vehicles Including Focus, Fusion for Door Latch Problem, Brake Fluid Leaks." https://www.usatoday.com/story/money/cars/2020/06/10/ford-recalls-2-5-m-vehicles-latch-brake-troubles/5338286002/. Retrieved August 14, 2020.

Lamprecht, J. L. (1992). *ISO 9000: Preparing for Registration*. Milwaukee, WI: Quality Press.

Liker, J. (2004). *The Toyota Way*. New York: McGraw Hill.

Linville, D. (February 24, 1992). "Exporting to the European Community." *Business America*. pp. 11–13.

MacLean, G. E. (1993). *Documenting Quality for ISO 9000 and Other Industry Standards*. Milwaukee, WI: Quality Press.

Madrigal, A. and R. Meyer. (May 21, 2020). "How Could the CDC Make That Mistake?" https://www.theatlantic.com/health/archive/2020/05/cdc-and-states-are-misreporting-covid-19-test-data-pennsylvania-georgia-texas/611935/. Retrieved on June 6, 2020.

Manganelli, R. and M. Klein. (1994). The *Reengineering Handbook: A Step-By-Step Guide to Business Transformation*. New York: AMACOM.

Masterson, P. (December 3, 2019a). "Recall Recap: The 5 Biggest Recalls in November 2019." https://www.cars.com/articles/recall-recap-the-5-biggest-recalls-in-november-2019-414302/. Retrieved on December 6, 2019.

Masterson, P. (January 7, 2019b). "The 10 Biggest Recalls in 2018." https://www.cars.com/articles/the-10-biggest-recalls-in-2018. Retrieved on December 6, 2019.

McNemar, Q. (1960). "At Random: Sense and Nonsense." *American Psychologist* 15, pp. 295–300.

Mills, C. (1989). *The Quality Audit: A Management Evaluation Tool.* Milwaukee, WI: Quality Press and New York: McGraw-Hill Publishing Company.

Montgomery, D. (2013). *Introduction to Statistical Quality Control* (7th ed. Chap. 8). New York: John Wiley and Sons.

Nehrer, A. (1967). "Probability Pyramiding, Rescaicli Irror and the Need for Independent Replication." *Psychological Record* 17(11), pp. 257–262.

NHTSA (National Highway Traffic Safety Administration). (December 2018). Report to Congress: "Vehicle Safety Recall Completion Rates Report. Biennial Report, #2 of 3." https://www.nhtsa.gov/sites/nhtsa.dot.gov/files/documents/18-3122_vehicle_safety_recall_completion_rates_report_to_congress-tag.pdf. Retrieved on April 15, 2020.

Niles, K. (2002). "Characterizing the Measurement Process." *iSixSigma Insights Newsletter* 3, p. 42.

Ortiz, C. and M. Park. (2010). *Visual Controls: Applying Visual Management to the Factory.* New York: Productivity Press.

Pande, P., R. Neuman and R. Cavanagh. (2000). The *Six Sigma Way.* New York: McGraw Hill.

Parsowith, B.S. (1995). *Fundamentals of Quality Auditing.* Milwaukee, WI: Quality Press.

Peach, R. W. (ed.). (1994). *The ISO 9000 Handbook.* 2nd ed. Burr Ridge, IL: Irwin Professional Publishing.

QIM. (November/December 1993). "ISO Companies Reap Big Benefits." *Quality in Manufacturing*, pp. 3–4.

Ramsey, L. (June 4, 2020). "Researchers Just Retracted a Massive Study on Whether a Common Malaria Pill Can Help Treat Coronavirus." https://www.businessinsider.com/lancet-retracts-observational-study-on-hydroxychloroquine-for-study-2020-6. Retrieved on June 7, 2020.

Robinson, C. B. (1992). *How to Make the Most of Every Audit: An Etiquette Handbook for Auditing.* Milwaukee, WI: Quality Press.

Russell, J. (ed). (2000). *The Quality Audit Handbook.* 2nd ed. Milwaukee, WI: Quality Press.

Russell, J. (ed.). (2005) *The ASQ Auditing Handbook.* 3rd ed. (formerly called *The Quality Audit Handbook: 2000*), ASQ Quality Press.

Russel, J. (June, 2006). "Process Auditing and Techniques." *Quality Progress.* http://asq.org/quality-progress/2006/06/standards-outlook/process-auditing-and-techniques.html

Sayle, A. J. (1988). *Management Audits.* Milwaukee, WI: Quality Press.

Scott, A. and Ajmera, A. (October 8, 2019). "GE to Freeze, Pre-pay Pensions to Save Up to $8 Billion, Cut Debt." https://www.msn.com/en-us/money/companies/ge-to-freeze-pre-pay-pensions-to-save-up-to-dollar8-billion-cut-debt/ar-AAIoGqB?ocid=spartandhp. Retrieved on May 21, 2020.

Smith, J. (December, 2019). "Quality Is Secondary: Mediocrity Seems to Have Become the New Standard." *Quality*, 58(13), p. 13.

Stamatis, D. (1996). *Documenting and Auditing for ISO 9000 and QS 9000: Tools for Ensuring Certification or Registration.* Chicago, IL: Irwin Professional Publishing.

Stamatis, D. (1997). *The Nuts and Bolts of Reengineering.* Red Bluff, CA: Paton Press.

Stamatis, D. (1998). *Advanced Quality Planning: A Commonsense Guide to AQP and APQP.* New York: Quality Resources.

Stamatis, D. (2002–2003). *Six Sigma and Beyond.* Vol. 1–7. Boca Raton, FL: CRC Press.

Stamatis, D. (2003). *Failure Mode and Effect Analysis (FMEA): From Theory to Execution.* 2nd ed. Rev. and expanded. Milwaukee, WI: ASQ Press.

Stamatis, D. (2010). *The OEE Primer: Understanding Overall Equipment Effectiveness, Reliability and Maintainability.* Boca Raton, FL: CRC Press.

Stamatis, D. (2014). *Introduction to Risk and Failures: Tools and Methodologies.* Boca Raton, FL: CRC Press.

Stamatis, D. (2015a). *The ASQ Pocket Guide to Failure Mode and Effect Analysis (FMEA).* Milwaukee, WI: ASQ Press.

Stamatis, D. (2015b). *Risk Management Using Failure Mode and Effect Analysis (FMEA).* Milwaukee, WI: ASQ Press.

Stamatis, D. (2016). *Quality Assurance: Applying Methodologies for Launching New Products, Services and Customer Satisfaction.* Chapter 16. Boca Raton, FL: CRC Press.

Stamatis, D. (2019a). *Risk Management Using Failure Mode and Effect Analysis (FMEA).* Milwaukee, WI: Quality Press.

Stamatis, D. (2019b). *Advanced Product Quality Planning: The Road to Success.* Boca Raton, FL: CRC Press.

Stamatis, D. (2020). *Engineering Ethics.* Morgan Hill, CA: Bookstand Publishing.

Stoop, E. (October 31, 2017). "Layered Process Audit Programs: A Fast-Track Strategy for Reducing Cost of Quality." https://www.qualitymag.com/articles/94334-layered-process-audit-programs-a-fast-track-strategy-for-reducing-cost-of-quality. Retrieved on July 15, 2020.

Stoop, E. (January 21, 2020). "Layered Process Audits to Close the Loop on Safety." https://www.beaconquality.com/blog/using-layered-process-audits-to-close-the-loop-on-safety. Retrieved on February 12, 2020. [According to the National Safety Council, workplace fatalities have risen 17% since 2009 after decades of steady improvement in occupational safety, outpacing workforce growth over that period].

Stojanovic, S. (October 11, 2017). "How to Establish Error-Proofing Process According to IATF 16949." https://advisera.com/16949academy/blog/2017/10/11/how-to-establish-an-error-proofing-process-according-to-iatf-16949/. Retrieved on July 15, 2020.

Tague, N. (2005). *The Quality Toolbox.* 2nd ed. Milwaukee, WI: Quality Press.

Thompson, B. (1988). "A Note about Significance Testing." *Measurement and Evaluation in Counseling and Development* 20, pp. 146–148.

Tukey, J. W. (1980). "We Need both Exploratory and Confirmatory Data." *American Statistician* 34, pp. 23–25.

VDA. (2016). *Quality Management in the Automotive Industry: Process Audit – Part 3.* 3rd rev. ed. (Auditor edition). Berlin, Germany: Verband der Automobilindustrie.

Ward, A. (March 2014). *Lean Product and Process Development.* 2nd ed. Cambridge, MA: Lean Enterprise Institute, p. 215.

Westfall, R. S. (1973). "Newton and the Fudge Factor." *Science* 179, pp. 751–758.

Wheeler, D. (2006). *EMP III: Evaluating the Measurement Process & Using Imperfect Data.* Knoxville, TN: SPC Press.

Wong, J. (December 20, 2017). "The Biggest and Noteworthy Auto Recalls of 2017: Tesla, Fiat Chrysler, Honda and Ford Were All Bitten by the Recall Bug." Retrieved on December 5, 2019. https://www.cnet.com/roadshow/news/big-and-noteworthy-auto-recalls-of-2017/. Retrieved on December 6, 2019.

SELECTED BIBLIOGRAPHY

Bureau of Business Practices. (1992). *ISO 9000: Handbook of Quality Standards and Compliance.* Needham Heights, MA: Allyn and Bacon.

Hagigh, S. (February 24, 1992). "Obtaining EC Product Approvals after 1992: What American Manufacturers Need to Know." *Business America*, pp 30–33.

ISO 10013:1995 *Guidelines for Developing Quality Manuals*. Milwaukee WI: ASQC.

ISO 1431:2013. *Environmental Management — Environmental Performance Evaluation — Guidelines*. Zurich, Switzerland: International Organization for Standardization.

ISO 27001:2013. *Information Security Standard*. Zurich, Switzerland: International Organization for Standardization.

Military Specification. (1956). *MIL-Q-9858A: Quality Control System Requirements*. Washington, DC: Superintendent of Documents.

Stamatis, D. (August 1992). "ISO 9000 Standards: Are They for Real?" *ESD Technology*, pp. 13–17.

Stamatis, D. (December 1995a). "QS-9000 Revisions: Not Far Enough?" *Quality Digest*, pp. 12–14.

Stamatis, D. (1995b). *QS-9000 the Automotive Standard*. Milwaukee, WI: Quality Press.

Stamatis, D. (November 1995c). "Toward an Automotive Quality Standard: Revisions of the QS-9000 Fall Short of Expectations." *Sensors: The Journal of Applied Sensing Technology* 12(11), pp. 4–5.

Stamatis, D. (1995d). *Understanding ISO 9000 and Implementing the Basics to Quality*. New York: Marcel Dekker.

Voehi, F., P. Jackson, and D. Ashton. (1994). *ISO 9000: An Implementation Guide for Small to Mid-Sized Business*. Defray, FL: Lucie Press.

Wilson, L.A. (1996). *Eight Step Process to Successful ISO 9000 Implementation: A Quality Management System Approach*. Milwaukee, WI: Quality Press.